LUNATICK ASTRONOMY

Research Into The Astronomical Interests
Of The Lunar Society Of Birmingham

Andrew P. B. Lound

**Odyssey D£
Publications
2009**

First Published in Great Britain by
Odyssey DL Publications
110 Sandringham Road, Birmingham,
West Midlands B42 1PQ

www.odyssey.dial.pipex.com

First Edition 2008
Second Edition 2009

A catalogue record for this book is
available from the British Library

ISBN 978-0-9561111-1-1

Cover Design: Andrew Lound

Typeset in Times 10pt
Printed in Great Britain by
Bookbinding Direct, Keele University
Staffordshire ST5 5BG

Dedicated To
My Mother

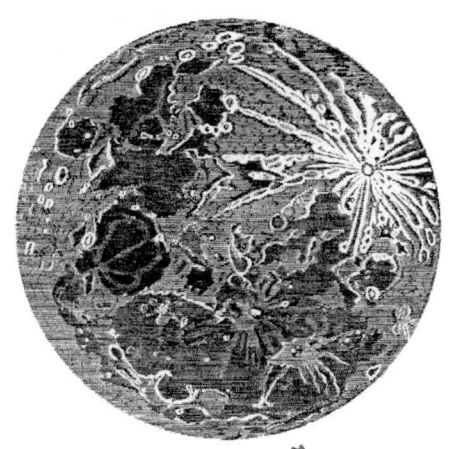

About the Author

Andrew Lound is a full time speaker, writer and broadcaster being the UK Coordinator for The Planetary Society, founder member of the Society for the History of Astronomy, Fellow of the British Interplanetary Society, and has held numerous positions with various astronomical societies. He has been a regular contributor on BBC Radio talking about space science since the late 1980s. He has also appeared in several television and film documentaries where he has played roles as diverse as the 17th century astronomer Jeremiah Horrocks, 2nd officer Lightoller of the *Titanic,* and Neil Armstrong. He has been active in promoting space science to the general public for over 30 years staging more than 2,300 exhibitions and lectures around the UK and abroad. He has developed the *Odyssey Class Dramatic Lecture;* this combines music, drama, and images with his detailed knowledge of space science both historical as well as contemporary. In 2006 working with the British Council he led a scientific and public awareness team to Libya to study a solar eclipse. He is also an expert on the famous liner *Titanic* and presents lectures and stages exhibitions on this subject as well. His research into the astronomical activities of the *Lunar Society* is one of several historical projects he is currently involved with the results of which are to be published in due course.

Contents

Introduction

The Lunar Society was one of the most influential gatherings of intellectuals in the eighteenth century. The group of friends met informally every month on a Monday evening closest to the full Moon (in order that they had some light to illuminate their way home). It is this singular astronomical connection from which the group of friends developed their collective noun – although they often referred to themselves as the 'Lunaticks' with the term 'Lunar Society' not appearing before 1776.[1] The Society met at various homes from around 1765[2] with Soho House (now in Birmingham) the home of Matthew Boulton being used more often than not. It is Matthew Boulton the Birmingham entrepreneur who pulled together the group of people who had common interests in science, nature, business, politics and dining to create a unique Think Tank. Matthew Boulton was the prime mover behind the moonlit gathering and most people today know him as the great manufacturer of Birmingham and the partner of James Watt. His friend and fellow Lunatick James Keir once observed that electricity and astronomy were at one time among Matthew Boulton's favourite amusements.[3] The question for myself with an interest in astronomy living only a few miles from Boulton's home is just how much of an interest in astronomy did Boulton and the other Lunar Society members have?

The vast amount of preserved material relating to Matthew Boulton and James Watt stored at Birmingham City Archives offers any researcher an almost too daunting a task to search the papers, notebooks, letters, drawings etc. Thus a narrow rather than wide-field search is necessary which has led to the familiar topics more associated with the members of the Lunar Society being selected for research, resulting in a number of books and papers being written concentrating on the development of steam engines, pottery, coinage, silver production, the abolition of slavery, and the development of modern industrial methods. Industrial historian Eric Robinson wrote two papers in 1957 exploring the Lunar Society's contribution to the developments of scientific instruments and suggested he would write a third about the connections with astronomy – a paper that appears to have never been written.[4] Robert E. Schofield in his book *'The Lunar Society of Birmingham'* refers in passing to the astronomical activities of the Lunar Society quoting James Keir. Thus it was time to re-examine the archive. The research was started in 1999 with the initial results being presented at a public event at Soho House Museum (Boulton's Home) on January 6th, 2002 which then formed the basis for a public exhibition at Soho House from March 8th to September 1st 2002.

The following is a full account of my research to date into Boulton's and the Lunar Society's interest in astronomy, a précis of which was first presented as a verbal paper which formed part of an event organized by the Society for the History of Astronomy (SHA) held at Soho House Museum on November 2nd, 2002 with a more detailed version being presented to SHA on June 1st at Soho House Museum, and later on August 30th, 2003 in Manchester. Details of Boulton's design of an observatory were shown at the Birmingham History Fair on June 15th 2008 and presented at the SHA Conference in Birmingham on October 4th, 2008. An Odyssey Class[TM] Dramatic Lecture entitled *'Lunatick Astronomy'* which is based on the research has been presented to numerous organizations across the UK and is available to any interested parties.

The write-up of the research has been produced on the urging of Professor Allan Chapman of Wadham College, Oxford, Professor Carl Chinn of the University of Birmingham, and Val Loggie former Curator of the Soho House Museum in order to assist those with an interest in the Lunar Society and to give wider public awareness of Matthew Boulton's astronomical interests. This publication contains additional material that is mentioned in the Dramatic Lecture aimed at a wider audience, placing together for the first time a complete ready reference to the Lunar Society's astronomical interests.

APBL Birmingham November 2008

Second Edition

The second edition includes information from sources that were not available prior to publication plus information from newly found documents. In particular the discovery of details of the telescope Boulton sold to Alexander Aubert; the connections between those who attended Ferguson's lectures in Kidderminster and the Lunar Society; a photograph of Number 11 The Square, the venue of Ferguson's 1761 & 1771 lectures in Birmingham; Ferguson's meridian line on St. Martin's Church; connection between Thomas Aris and Benjamin Martin's lecture tour of the midlands; James Keir's interest in the Scottish glass trade; William Herschel's visits to Soho House and the Midlands in 1792 and 1793; Herschel's evidence in the Steam Engine Patent court hearing; the nature of itinerant lecturers touring Birmingham in the 18th century; additional information for those not acquainted with the Lunar Society or Birmingham.
This edition has been reformatted.

APBL Birmingham May 2009

1. A Town Of Practical Science

The growth of Birmingham from a rural town to a major industrial city has been well documented with numerous books and papers being published, covering everything from the social life to the political life, individual companies to public houses. Within this wealth of material the intricate details of an urban area speaks to us as if the people of the various eras were still with us. John Leland (1502-1552) had, in 1538 visited Birmingham riding across the bridge over the River Rea entering into what was a market town. On either side of the road he saw numerous open smiths, inside the largest of which four men worked, each taking a turn to strike a hammer on the anvil. With the air filled with hammering, the smoke rising and the red and orange glow from the fires in the smiths, it resembled a scene from a mythical world.[1]

Birmingham was a gild free town, thus it was a place where anyone with a skill, a talent or the drive to succeed could make a go of it. The emphasis on the 'can do' approach attracted men and women from all parts of the country and consequently the character of the town followed this individualistic approach. It was a town where people did not always conform to the standard national average. Yet during the civil war Birmingham was like other industrial towns, by being against King Charles I (1600-1649) with particular opposition to his religious policies. In 1642 just prior to the Battle of Edgehill, King Charles was staying at Aston Hall as a guest of Sir Thomas Holte (1571-1654). Holte found himself at odds with the population who harried Charles' baggage train capturing his bodyguard and handing them over to Cromwell's men. They also prevented weapons reaching Royalist forces, ensuring they reached the Parliamentarians instead. A year later Prince Rupert (1619-1682) made Birmingham pay a heavy price in money and blood for such disloyalty with a particularly savage sacking of the town. Birmingham suffered in the years following the civil war with a slump in trade and then plague ravaged the area. It wasn't until the last twenty years of the 17th century that Birmingham began to prosper, with a rejuvenated iron industry, gun trade and of course the lack of a gild to shackle people. By the 18th century buckles, buttons and a variety of household metal goods were being produced in the town which grew in population and wealth.[2] With all of this came the need for better education – for those who could afford it. King Edward's Grammar school had been founded in 1547 and continued to develop with new buildings constructed in 1707. The rise of the nonconformist movement which encouraged education had an effect as well, the town was seen by many as a safe refuge for reformers of all kinds. The individualistic nature of Birmingham rose out of Prince Rupert's charred wood.

Many other towns began to form local societies that would meet at coffee houses or in rooms above inns. The records do not show a mass of societies in Birmingham although small private groups may have existed and their details have been lost. The Lunar Society is of course an example of a private group of friends meeting with common interests and it is only through the letters of the members and visitors to the Society that we have an idea of what occurred. In London the coffee houses produced a large number of political, literary and scientific societies; the towns which had a strong academic tradition such as Oxford, Cambridge, Edinburgh and Glasgow also had people coming together to discuss common interests. Birmingham was a town of artisans, many of an individualistic nature who tended to do their own thing. These people did have an interest in the world around them, in science, art and politics, but they took a pragmatic approach and were interested in subjects that could be of practical use. This practical approach manifested itself in apprenticeships, working with fathers in their trade and by simply setting up on their own. Further education as we would now call it was in great demand mainly if the subjects were of practical use that could not otherwise be learned on the job.

The source of this education was through the itinerant lecturers, groups of people who toured the country presenting a series of lectures on a select subject often with impressive demonstrations using a variety of apparatus such as orreries, models and even living creatures. Birmingham's townsfolk were treated to many such travelling lectures, and attendances seem to be high as many lecturers made return visits (see appendix A).

In August 1742 an exhibition of optical equipment was on display at the *Wheatsheaf* inn 8½ The Bull Ring as advertised in Aris' Gazette.

> The Most ingenious and unparallel'd OPTICAL INSTRUMENTS, being a
> specimen of the works of the most celebrated Sir ISAAC NEWTON.........
> To be seen by one person or more all Day long till ten at Night.
> To the Gentry 1s to young Persons, &c 6d [3]

Thomas Aris (d.1761) was a local bookseller and newspaper proprietor from London who set up a newspaper in 1741, although he thought he would be the first a Mr. Walker had beaten him to it and the two newspapers vied each other until common sense prevailed and they united. Aris' Gazette was a big boost for Birmingham and it enabled itinerant lecturers to advertise their presence in the town and to possibly raise subscriptions for additional lectures.

Three months after the optical exhibition the town was visited by another touring attraction *'The Microcosm'*, by H. Bridges described as a world in miniature it featured two clockwork planetariums designed to demonstrate the true movements of the planets. The exhibition was located at the *Old Cross* in High Street with an admission of one shilling.[4]

It was another sixteen months before the next science entertainment arrived, and this time it was a set of formal lectures – probably more to the liking of a practical audience – held at Mr. Beckett's in New Street; Francis Midon presented eight philosophical lectures, seven of which were to be upon air and the eighth on attraction and electricity. The advertisement in Aris' paints a wonderful picture:

> An eighth lecture on ATTRACTION and ELECTRICITY , containing all the most
> curious Experiments upon the Subject and particularly the following: A Boy, with
> his clothes on being suspended by two Silken Cords, and a Glass Tube brought
> near the Soles of his Shoes, his Face and Hands will be endued with a strong
> attractive Virtue.....copious and vivid Emanations of Light will be seen and heard
> to issue with a snapping Noise from the ends of his fingers.[5]

The advertisement also makes it clear that common terms would be used during the lecture to make it 'familiar to persons of both sexes,' with a fee of one Guinea. The lectures must have been a great success as Midon presented lectures for another three months.[6]

Midon had probably been an itinerant lecturer for some time with his standard set of seven lectures on air, his eighth on electricity illustrates the great interest by scholars, amateur scientists and the public in what seemed to be a magical effect. Lecturers began to tour with electrical apparatus that would entice the public to participate. Birmingham people like all others were just as curious as they attended the demonstrations of Thomas Yeoman (1709/10–1781), and Mr. Smith – demonstrations that were as much entertainment as science, but then if you can enthral your audience they will learn better.[7]

Birmingham found itself playing host to new lecturers at different locations, the inns of the town may well have been in competition with each other to provide an attraction or perhaps some of these inns had their own little societies; either way it was a great opportunity to learn science. An astronomy and navigation course by Mr. S. Hayes began in Moor Street in August 1748;[8] on the 8th May the following year J. Arden a mathematics teacher from Derby began his first series of lectures on *Natural and Experimental Philosophy*, he was known to Thomas Aris who had printed a series of science problems set by Arden in 1742;[9] the *Microcosm* returned from July through August.[10] The *London Apprentice* had played host to *Mr. Baker the Modern Living Colossus* a very tall man who also attended the local fair[11] now it hosted a demonstration of the Microscope to the 'Philosophical Gentlemen' of the Town.[12] 1755 through 1756 was a busy period with William Griffiss delivering his experimental lectures in chemistry at Mr. Packwood's in April and May 1755, and then at Mr. Hambleton's the upholsterer at the *Old Cross* in High Street in August, September and October 1756.[13] Griffiss seems to have been doing well, he had presented his lectures in Dudley and had added astronomy to his programme:

1 The principles of Sir Isaac Newton's Philosophy illustrated & confirm'd.
2 Mechanicks illustrated by a variety of models.
3 Hydrostaticks & Hydraulicks or the Art of raising water.
4 Pneumaticks or the Surprizing Properties of Air.
5 Chymistry shewing its application to the improvements of Trade in all Arts.
6 Opticks and how short sighted and old eyes may be helped.
7 Fortification illustrated by a large model of a fortified town.
8 & 9 Geography & Astronomy illustrated by Globes, Planetariums Cometariums and the Grand Orrery much improved. [14]

One young Birmingham man was fascinated by Electricity and Astronomy who may well have attended some of these lectures; he was the son a buckle and button maker by the name of Matthew Boulton.

2. Lunar Society's Astronomical Beginnings

Matthew Boulton was born in Birmingham in 1728 the son of a successful buckle and button maker with a manufactory in Birmingham located on the road to Wolverhampton known as Snow Hill. When Matthew came of age he entered the family business as a partner and showed great ability in many subjects having a most enquiring mind. He developed new buckles and one of his spare time pursuits was in making thermometers. In around 1756 he married Mary Robinson (1727-1759) the daughter of Luke Robinson (1683-1749) a wealthy mercer of Lichfield; she was one of two sisters who were co-heiresses along with their brother to quite a fortune. Sadly the marriage did not last long as Mary died in 1759. It was a difficult time for Matthew as in the same year his father died suddenly, and thus he inherited the business which he set about developing. Boulton wanted to be a great silversmith, and as a salesman and motivator he was supreme. He entered into partnership with John Fothergill (1730-1782) in 1762 forming the firm of *Boulton and Fothergill* located at the newly built Soho Manufactory – which at the time was the world's largest and most modern facility producing a wide range of quality articles such as silverware, ormolu, clocks, watch chains, etc.

Fig. 1 Matthew Boulton Birmingham entrepreneur in early life. (Private Collection)

The setting up of the manufactory was an expensive undertaking thus the need for a partner although in 1760 Boulton had a windfall when he married for a second time – Ann Robinson (1733-1783) the sister of his first wife thus gaining the second half of the inheritance of the Robinson sisters![1]

His business along with his first and second wife's fortune meant that he was able to afford to indulge in a number of pastimes. It was through a combination of business, his pastimes and personal interests that he began making contacts around the country and abroad. Some of these contacts would become life-long friends and would form that unique group known today as the *Lunar Society*. The names of the members developed over the years and reads like a who's who of 18th century intelligentsia: Matthew Boulton (1728-1809), Erasmus Darwin (1731-1802), Thomas Day (1748-1789), Richard Lovell Edgeworth (1744-1817), Samuel Galton Jr. (1753-1832), Robert Augustus Johnson (1745-1799), James Keir (1735-1820), Joseph Priestley (1733-1804), William Small (1734-1775), Jonathan Stokes (1755-1831), James Watt (1736-1819), Josiah Wedgwood (1730-1795), John Whitehurst (1713-1788) and William Withering (1741-1799). Then there were the guests to the moonlit dinners such as Benjamin Franklin (1706-1790), William Herschel (1738-1822), James Lind (1736-1812), Jean André de Luc (1727-1817), James Ferguson (1710-1776), John Michell (1724-1793), William Murdoch (1754-1839), Horatio Nelson (1758-1805), Joseph Black (1728-1799) and John Robison (1739-1805) to name but a few. In an atmosphere of wine and food, serious science was discussed and in many cases experiments planned and carried out.

Matthew Boulton recorded many of his thoughts, ideas, plans, calculations and engineering designs in a number of notebooks [2] one of which contains a mixture of notes that date from 1751 to 1759. It is in this book that we first see written evidence of Matthew Boulton's interest in astronomy (and meteorology). On October 20th 1759 he notes a table of meteorological observations made at Derby and Buxton with his friend John Whitehurst, and has added "Ball of Fire" referring to an observation of a meteor on October 20th.[3]

His interest in meteorology appears to have led to an interest in the heavens and he began to study the subject and learn the first principles of astronomy. Like all those setting out in the pursuit of the subject Kepler's Laws of Planetary Motion is required learning along with basic time calculations for Solar and Sidereal time. Boulton like any other student of astronomy duly writes down the details in his notebook in order to remember them.

"Planets or comets w[de] move in elipsis describe equal areas in equal times.

The squares of the planits period round the Sun are as the cubes of their distances from the Sun & the force of the attraction of the Sun upon them is inversely as the squares of their distances from the Sun.
The same law takes place in the satellites moving around their primary planets.

	Hours/min	Seconds	Thirds	
Solar Day is	24			
Sidereal Day is	23 = 56	= 3	= 27	"[4]

It is perhaps not surprising that Boulton's interest in astronomy should begin in the late 1750s, during this time the talk of those interested in science was of the possible reappearance of 'Halley's Comet', the appearance of which had been predicted by Edmund Halley (1656-1742) utilizing the work of Isaac

Newton (1642-1727). The comet was due to make an appearance around 1758-9 and if it did appear then it would vindicate the work of Newton and it would set his work *'Principia'* as *the* text on the movement of the heavens – as well as an application to engineering.

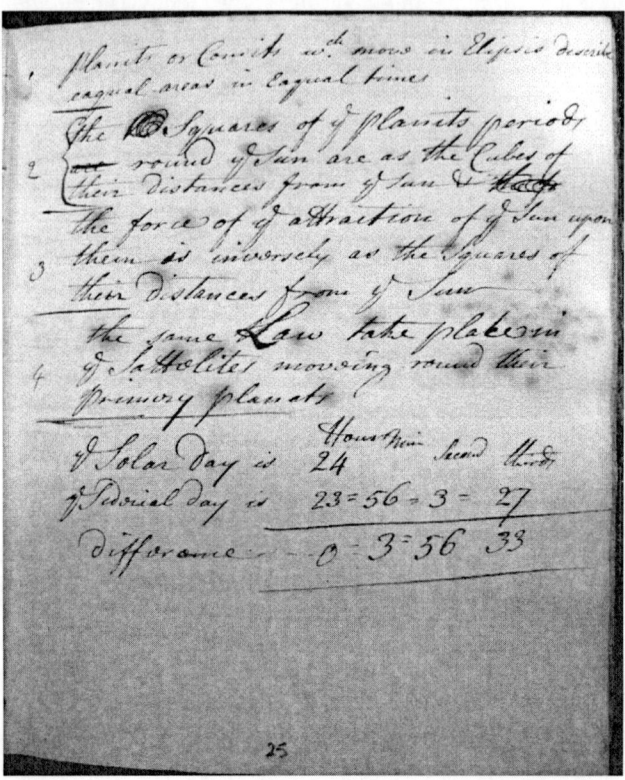

Fig. 2   One of Matthew Boulton's Notebooks showing his notes of Kepler's Laws.
(Birmingham City Archives)

The possible reappearance of the comet was hotly debated and promoted to a wider public through itinerant lecturers. Benjamin Martin (1704-1782) was one such lecturer who visited Birmingham in 1747 to present at least two series of twelve philosophical lectures; these were held at Mr. Taylor's Great Room in the Cherry Orchard (the present Cherry Street off Temple Row in Birmingham City Centre is close to the venue). Martin was one of the great promoters of Halley's Comet's return and in his advertisement he emphasises his new demonstration equipment including a new orrery and a *Cometarium*. Three years later he gave another series of twelve lectures at Mr. Packwood's at the *Bowling Green* in Cherry Orchard. All of these lectures were supported by Thomas Aris.[5] Matthew Boulton would have been only 19 when Martin gave his first set of lectures and it would have been an excellent opportunity to stimulate a young mind. In 1755 Martin began publishing the *'General Magazine of Arts and Sciences'* in which he included details of the forthcoming comet. Martin advertised a plan of his periodical in Aris' Gazette which was free from Thomas Aris as a taster for the full magazine that was to be published on 1st February.[6]

With all of the visiting lecturers demonstrating their apparatus many in Birmingham should have been prepared for the return of Halley's Comet. On Christmas night 1758 a German amateur astronomer Joannes G. Palitzsch became the first to spot the comet on its predicted return, resulting in Newton's work becoming the prime text on planetary motion.

James Ferguson a Scottish astronomer, artist and clock maker published a book entitled '*Astronomy Explained Upon Sir Isaac Newton's Principles*' that clearly explained Newton's ideas. It became a best seller, its style made it readily accessible to anyone who understood geometry and general mathematics. This book was to inspire and influence many into studying astronomy. The transit of Venus in 1761 was another astronomical event that was causing great interest. This event only occurs twice in a period of over 100 years with the first recorded event taking place in 1639 by Jeremiah Horrocks (1618-1641). Edmund Halley had done much work to prepare astronomers to observe the predicted 1761 event which once accurately observed would provide a calculation of the distance between the earth and the Sun – the so called *Astronomical Unit* which could be used as a baseline for measuring the universe. Ferguson updated his book to include details of the transit and engaged on a lecture tour with a special orrery to demonstrate the event. Ferguson was acquainted with John Whitehurst, Erasmus Darwin and Benjamin Franklin all of whom were part of Boulton's growing circle of friends. Ferguson was invited by 'friends' to visit Birmingham in August 1761 to present a series of lectures on '*Experimental Philosophy*'. Ferguson had been lecturing regularly in London but began to increase his range of activity so long as a sufficient number of subscribers could be found in a particular town. He then advertised his presence in the town hoping to attract subscribers to additional lectures. Benjamin Martin had stopped touring in 1756 after he settled in London and opened a shop selling scientific instruments. Ferguson's invitation may have been the result of Martin's decision to stop touring. Like Martin, Ferguson's series of lectures was generally in twelve parts and included demonstrations using scientific apparatus built by himself– the kind of thing that appealed to Boulton and many if not all of his friends. Who Ferguson's friends were who did the inviting is not listed however for the period it can be seen that Boulton, Darwin, and Whitehurst were close friends by this time and may have been initial subscribers. Birmingham's famous historian and bookseller William Hutton (1723-1815) may well have been one of these friends; he observed the 1761 transit of Venus from Aston.[7]

Ferguson's lectures were held at Mr. Sawyer's Great Room (also called the *Assembly Room*), at number 11 *The Square*.[8] The Square was one of Birmingham's most fashionable areas and over the years many notable people resided or visited there including Dr. Samuel Johnson (1709-1784) whose friend Edmund Hector resided at number 1 – known as 'Hector's House'. William Sawyer had been resident at numbers 10 and 11 since 1740, he taught dance, with his mother and sister running a school. The Assembly Room was used for many functions including concerts, musical recitals and lectures; the location may have been chosen by William Hutton whose daughter Catherine (1756-1846) attended the Sawyers' school.[9] The programme of lectures started at 4 o'clock each day (Sundays excepted) and ran in sequence as follows:

1. Properties of Matter, centrifugal, centripetal forces, laws by which the planets move, tides, earth's axis and orbital motions.
2. The five mechanical powers (levers, wheel and axle, wedge, pulleys, screw) shown singularly and in combination; thermometer & pyrometer.
3. Wheel carriages, water mills, hand mills, pile driver, all explained by working models.
4. Specific gravity and detection of counterfeits, glass models of sucking & forcing pumps, models of the Persian wheel and a quadruple pump mill.
5. Nature and properties of air, experiments with the air pump on the weight of the air.
6. Experiments with the air pump on the elasticity of the air.
7. Use of the globes, Dr. Long's glass sphere, equation of time.
8. Dialling in general, construction of different kinds of dials.
9. The solar system demonstrated by the orrery and cometarium.
10. The seasons, velocity of light, longitude by the eclipses of Jupiter's moons.
11. The moon's rotation on her axis, phases, harvest moon.
12. Various demonstrations with the orrery & eclipsarion, the year of the Crucifixion determined astronomically.[10]

Fig. 3 James Ferguson Scottish astronomer and clock maker who presented lectures in Birmingham in 1761 & 1771. (From a print in the Lound Collection)

The initial set of twelve had been pre-booked by subscription of a Guinea per series. The lectures were so successful that Ferguson remained in Birmingham for 3 months presenting four series of lectures.[11] He was well received in the town and in between lectures he visited local manufactories and this is highly likely to have included Boulton's at Snow Hill.[12] The style of Ferguson's presentations especially with the demonstrations would have been a magnificent experience. For the likes of Boulton it must have been thought provoking and highly stimulating. Soon after the first set of lectures Boulton bought Ferguson's book (1757 edition) purchased from *Pearson and Aris* for eighteen shillings.[13] Prior to this book he was probably working with David Gregory's *'The Elements of Physical and Geometrical Astronomy'* a two volume set the second edition of which was published in 1726. Whether this originally belonged to his father I do not know.[14]

Fig. 4 The Square in Birmingham, it was at number 11 where James Ferguson presented a series of lectures in 1761 & 1771.
(From a drawing by W. Westly 1732 Birmingham City Archives)

Boulton's interest in astronomy naturally led him to an interest in optics, in Notebook 4 (late 1750s early 1760s) he lists the optician Peter Dollond (1730-1820) the son of John Dollond (1706-1761) as a friend,[15] purchasing optical materials from him including wedges of glass to explain refraction for £1 11/6d.[16] It was in this period that Boulton began making many connections probably just as much for his personal interests as well as for business purposes. The Reverend John Michell of Queen's College Cambridge became an important (as with many of the other Lunaticks) contact. He encouraged Boulton in his electricity and astronomy pursuits.

Boulton's interest in electricity would have been greatly aided by his friendship with Dr. Benjamin Franklin who Michell introduced to Boulton (and to Erasmus Darwin) in 1758.[17] In turn Franklin introduced Boulton to Dr. William

Small who had recently returned to Britain from Virginia in the United States where he had been Professor of Natural Philosophy and Mathematics having as a pupil the future United States President, Thomas Jefferson (1743-1826).[18] Small had been purchasing scientific instruments to equip the William and Mary College at Williamsburg and these included instruments by Peter Dollond and Edward Nairne (1726-1806). Small was a Scotsman who matriculated at Aberdeen in around 1750 achieving his degree in 1755. He came to Birmingham in 1766 and worked with Dr. John Ash (1722-1798) helping to found Birmingham's General Hospital. Michell of course seems to know everyone and is connecting many people together, interestingly he hung out at the same tavern in London as Boulton and his friends – the *Crown And Anchor* located in the Strand.[19] Boulton was there often discussing business and in particular with politicians regarding the setting up of an Assay office in Birmingham and Sheffield in 1773.[20]

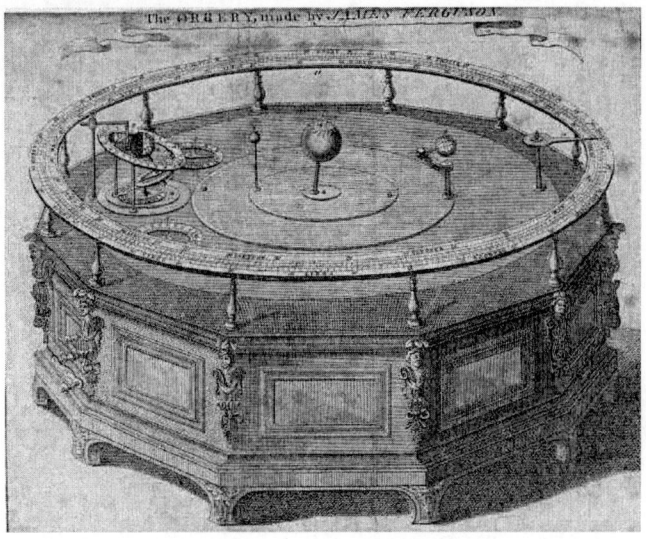

Fig. 5 Ferguson's Orrery with which he demonstrated the movements of the planets during his lectures. (From a print in the Lound Collection)

Although William Small worked in Birmingham as a medical doctor he continued his deep interest in science especially with optics entering into correspondence with Peter Dollond in 1765.[21] A much longer series of correspondence with James Watt in Glasgow (then an instrument maker/repairer) on many subjects including the improvement of telescopes, microscopes and micrometers demonstrates his depth of interest in the subject. Watt and Small were interested in the possible improvements in optics and the benefits of flint glass in lens making, and both were keen to begin experimenting.

James Watt was the fourth son of James Watt (senior) (1698-1782) of Greenock a ship's chandler whose father Thomas (1639/42-1734) was a teacher of mathematics and a great influence on the younger James. James went to London to study scientific instrument making working at one point with the well known telescope maker James Short (1710-1768) who had a shop at Surrey Street, in the Strand. Watt soon developed a reputation as an excellent instrument maker particularly with quadrants, microscopes, sea compasses as well as musical instruments![22] His reputation as an excellent cleaner and repairer of astronomical instruments followed his work in 1756 on a large set of astronomical instruments bequeathed to Glasgow University by Alexander MacFarlane (d.1755). MacFarlane had been a merchant living in Jamaica and had set up an astronomical observatory after obtaining instruments from Colin Campbell. These instruments included a 12 inch Gregorian telescope by Short, an astronomical clock by George Graham (1673-1751) and a 4 foot brass arch by Jeramiah Sisson (1736-1788).[23] These had suffered during the journey from Jamaica to England and Watt was given the task by Dr. Robert Dick to collect and return them to working order.[24] Watt was paid five pounds for the work[25] which was of such good order it brought him to the attention of John Robison also of Glasgow University. Robison was interested in expanding his and Glasgow University's knowledge in astronomy and was impressed with Watt's knowledge of science and instrument making. It was Robison who would later give a model of a Newcomen steam engine to the instrument maker for him to make repairs, which would lead to bigger things!

Fig. 6 Alexander MacFarlane Scottish merchant who lived in Jamaica. He bequeathed a large set of astronomical instruments to Glasgow University which was cleaned and repaired by James Watt. (Glasgow University)

Alexander Wilson (1714-1786) was also impressed by Watt, Wilson worked with Glasgow University as a surveyor and he took an interest in the MacFarlane Collection. He would eventually be appointed as Professor of Practical Astronomy and Observer at the University – a position created for him with the observatory being equipped with the MacFarlane collection (the observatory was often referred to as the *MacFarlane Observatory*). Wilson and Watt worked together on surveying the Clyde and his influence on Watt and the improvement of scientific instruments was profound. Wilson entered into a partnership with Watt and Joseph Black selling quality instruments with Watt noting an astronomical clock they sold to Messrs. Wilson & Cosan for £7 10/- .[26]

Watt had been running a business repairing and selling scientific instruments mainly for the shipping industry. His first sale was an object glass for a 3 foot telescope sold at one shilling and sixpence.[27] From an edited list of items he sold we can see the scope of his work (Table 1).

Date		Cost		
1757		£	s	d
In stock	Hadley's Quadrant Compleat	1	17	
Ditto	5 sets of glasses for Hadley's Quadrant	1	4	
Ditto	2 sets of glasses for 3ft sliding telescope		5	
Ditto	5 sets of glasses for common 3ft telescope		9	
Ditto	3 sets of glasses for 2ft, 3 sets glasses for 1½ ft telescope		9	
26 Jan 1757	Sold Object glass for a 3ft telescope		1	6
	Sold Hadley Quadrant		2	3
2 Feb 1757	Sold a glass for a 3ft telescope		1	6
13 Feb 1757	Sold Hadley Quadrant		2	3
	Silvered glasses		1	
6 April 1757	Sold a glass for a 3ft telescope		1	6
9 May 1757	Mr. Richard Hill mending Telescope		1	6
27 Jan 1758	Glasgow College Mending telescope		3	
	Glasgow College Mr.McLeish Quadrant glass		5	
20 April 1758	John Robison Eyeglass of microscope			8
	John Milne 10 Telescope Mirrors on return			
3 April 1759	Bought copper & tin for speculums		10	
12 Sep 1759	Mending telescope		5	
21 Sep 1759	Sold John Muirhead telescope		2	4
30 Sep 1759	Sold Mahogany box for 3 ft telescope		14	

Table 1. Edited list of work undertaken in James Watt's shop taken from a copy of James Watt's Scrapbook 1757-1760. (Birmingham City Archives).

James Watt's waste book reveals a great deal about his work in optics, there are many entries where his father bought Hadley's Quadrants for his own chandler's shop. The quadrant was used to measure angles between objects; in astronomy this was between celestial bodies, in navigation the angle between a star or the sun and the horizon would be measured. The most well known of the quadrants was one invented by John Hadley (1682–1744) in 1732 (technically it is an octant as its curve is an eighth of a circle rather that a quarter of a circle however it was still referred to as a quadrant by sailors). Hadley's instrument is the precursor to the modern sextant and in the eighteenth century good makers of Hadley's Quadrant would find a large market as every ship's navigator needed one. Ship's chandlers fitted out vessels with virtually everything that was needed for a voyage and they would source navigational equipment from a variety of instrument makers. James Watt was one such instrument maker, one who was noted for his expertise on the Hadley's Quadrant.

Watt had already corresponded with James Lind (the son of George Lind the Lord Provost of Edinburgh, and cousin to Dr. James Lind (1716-1794) famous for finding the cause of Scurvy) on the subject of optics and scientific instruments. In 1765 Lind had purchased a 'noble sextant' from his friend the instrument maker Jesse Ramsden (1735-1800) with which he was delighted. The instrument was only six inches in radius yet it was divided to half seconds and had many 'pretty contrivances about it.'[28] The detail and accuracy of the instrument appealed to James Watt whose life would be spent producing instruments and parts thereof as accurately as possible. Lind goes on to say:

> "…Dollond has made great improvement on his glasses on making the making of them take quite high a charge as to equal reflector of the same length and to exceed them in brightness as of those I have (illegible) got one of 3 foot and a half is £21 – 0 – 0 mine is 2 feet and a half what it will cost I cannot tell. The opera glass of this construction on some nights ago one of five inches in length I could distinctly see Jupiter's satellites. Rare news for the astronomers." [29]

Lind informed Watt in the same letter that Ramsden had improved the astronomical quadrant and the solar microscope. A few months later Lind writes again saying he wants an equatorial stand for one of Dollond's latest telescopes but would like to consult with Watt first.[30] Having consulted with Watt, in October 1768 he wrote with his decision:

> "… I am getting a paralactic stand for one of Dollond's 36 inch telescopes, Miller the turner's son is my operator he has been many years away in London." [31]

Watt had a lifelong friendship with Lind who even sent his PhD thesis to Watt for him to comment on before submission. Lind eventually graduated as a medical doctor and moved to Windsor to become Physician to George III (1738-1820) – a connection that would benefit Watt in the years to come.

Watt continued to repair telescopes for the University and for private individuals (see Figs 9, 10, 11), thus it was natural for William Small to correspond with him on matters relating to the improvement of optics. Watt wrote to Small on January 31st, 1770:

> "I wish you saw the paper I mention on optics, you know I am no optician and this man is so entrenched with algebra that there is no coming near him but some metaphysical reasoning on the subject weighs with me. I propose when you and I meet to spend some of our leisure time on trying some experiments etc on glass for Dollonds."[32]

Watt was in need of a telescope but seemed reluctant to pay a high price for a Dollond lensed instrument. Small wrote to Watt in February 1771:

> "Mr. Boulton has got a workman who makes achromatic object glasses better than Dollond."[33]

The workman in question was a Frenchman by the name of Alexandre Tournant (c.1730-1792). Lenses produced by Dollond were expensive – although well within the price range of Boulton. However it appears he wanted to have a number of telescopes of various sizes for in 1770 he employed Alexandre Tournant who had a reputation for making lenses of superior quality.

Fig. 7 James Watt scientific instrument maker who is better known for his work on the steam engine. His early career was concerned with navigational and astronomical instruments. (From a print in the Lound Collection)

Tournant had previously worked for Francoise Thomas Germain (1726-1791) and afterward at the Academy of Sciences in Berlin; he began work at Soho Manufactory as a turner and chaser, and as Boulton's personal telescope maker paying him two Guineas a week plus a house for himself, his wife and some children.[34] Tournant was not simply a maker of lenses he also assisted Boulton in other aspects of manufacturing such as creating a formula for Red Gilding Wax.[35] Boulton recommended Tournant to (Charles Lennox) the Duke of Richmond (1735-1806) who in 1770 wanted some telescopes to carry out a number of experiments of his own; Boulton wrote:

> "I have now a man (Tournant) that I can recommend and whose honesty I will guarantee He is a man of good sense and good manners, he can draw & model ornaments,... he can fit and file accurate enough for the best mathematical instruments he is a very good turner either round or rose work ...made for himself a very curious lathe he grinds optical glass very true and hath made some exceeding good acromatic telescopes indeed I know not of a man in England so likely for your Grace's purposes".[36]

The lathe may well be the copying lathe for which Tournant is famous along with a lens polishing machine that may have been the reason why Small and Boulton considered his lenses to be of superior quality. It was for these machines as well as his work in optics that Tournant was awarded 5,000 livres from the *Bureau de Consultation* in France.[37] As can be read in Boulton's letter he thinks incredibly highly of Tournant, introducing him to Jesse Ramsden.[38] Tournant also began corresponding with Watt and Small on the production of lenses and micrometers.

Fig. 8 William Small he was a great influence on Boulton and corresponded with Watt on optics and steam engines. (Birmingham City Archives)

Watt was interested in purchasing a telescope and was wondering if Tournant would sell him one although Small informed Watt that Tournant was in the private employ of Boulton.[39] Tournant was also about to move on which he did in 1772 although he remained in contact with Watt, Small and Boulton on the subject of optics and micrometers and it was his high regard for the flint glass produced by William Parker that led another Lunatick Joseph Priestley to obtain his glass instruments from Parker.[40] Yet again however it was Matthew Boulton's connections that had brought Parker to the attention of Tournant. Boulton's interest in all things scientific had led him to Parker's flint glass work and through his agent John Wyatt he purchased material from Parker for Tournant.

Tournant and his family remained in contact with Boulton and Watt for many years, and when Tournant died in 1792 (soon after receiving his award from the Bureau de Consultation) Boulton helped set up Tournant's wife and daughter in Dublin running an instrument making business *Butler & Tournant* of 31 Grafton Street – no doubt the proceeds from the award enabled the business to be viable.

Watt and Small's correspondence regarding optics focuses heavily on Dollond's patent for the achromatic lens. Small and Watt were not pleased with the patent as they felt that it prevented further experimentation in optics apart from the fact that the principle mentioned in the patent was already known to many from the work of Chester Moor Hall (c.1703-1771).

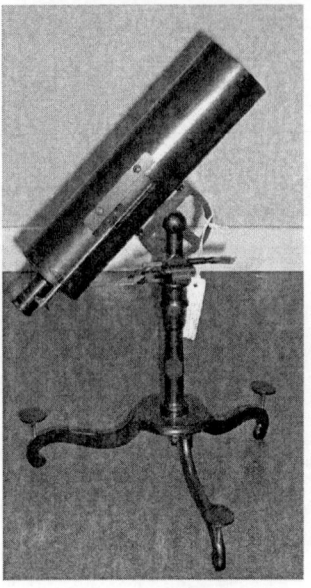

*Fig. 9 Telescope repaired by
James Watt.
(Located at Hunterian Museum,
Glasgow University)*

*Fig. 10 Telescope repaired by
James Watt. (Located at Hunterian
Museum, Glasgow University)*

*Fig. 11 Detail of telescope repaired by James Watt.
(Located at Hunterian Museum, Glasgow University)*

They were not alone in opposing the patent although the issue has more to do with John Dollond's son Peter (a friend of Matthew Boulton). John Dollond had shown little concern for competitors producing optical equipment using his technique but on his death in 1761 Peter, his eldest son took over the business and let it be known that he required a royalty on all sales or else he would prosecute. A petition against the patent was submitted by London opticians and instrument makers - a petition it seems through their correspondence was supported by Small and Watt – the petitioners included Jesse Ramsden and Benjamin Martin. It came to nothing, and Dollond was successful in the courts against Joseph Champneys of Cornhill in 1766 which put an end to opposition, resulting in many opticians going bankrupt.

Watt's viewpoint on the prevention of experimentation is interesting given that in the years to come his patent for the steam engine with a separate condenser would be criticized for exactly the same reason. It may also however have been a matter of cost that concerned Watt. Regular two-foot telescopes from his shop sold at ten shillings [41] Achromatic telescopes had to include a royalty payment to Dollond which raised the cost of telescopes to a Guinea per foot length.[42] Watt looked at supplying reflecting telescopes and a variety other instruments including orreries calculating how much profit was to be made on them; using Benjamin Martin's catalogue and price list as a guide he noted the following:

Large Orreries	£150	profit - £10 10s
Senex Globes 28 inch diameter in Mahogany frames £35		profit - £ 4 4s
A Three foot reflecting telescope mounted on a Brass foot		
	£18 18s	profit - £ 2 2s
A Two foot ditto	£11 11s	profit - £ 1 1s[43]

Dollond's patent expired in April 1772 with Watt and Small persevering with optical experimentation as Small wrote to Watt in 1773:

"I am attempting the improvement of telescopes and still more anxiously of Microscopes, because the present Microscopes deceive their users; but I find it very difficult to procure good lenses. Could you make an achromatic lens of ½ an inch focal distance? "[44]

Watt replied in March:

"I have invented two problems for clearing the observed distance of the Moon from a star of the effects of refraction and parallax; one trigonometrical, by Mercator's sailing, - the other instrumental, by a sector having a line of chords on each limb and a moveable portion of a circle of the same radius, which, if of three feet, the problem may be solved to ten seconds."[45]

James Watt is clearly making improvements, in his journal in October 1773 he notes:

"8th to the 24th (October 1773) trying experiments upon telescopes, found that object glass 15 inch could bear an eye glass of ¾ inch easily – invented a new micrometer consisting of a telescope with an object glass a field glass and an eye glass with two parallel cross hairs in the focus..."[46]

Small wrote to Watt a few months later:

"I rejoice in all of your improvements, but have many optical difficulties, that lessen my confidence in observations made with the most accurately divided instruments..."[47]

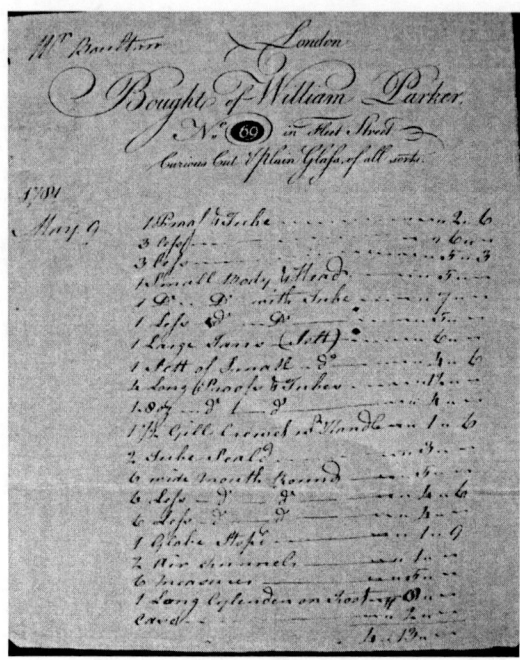

Fig. 12 Bill to Matthew Boulton from William Parker, Boulton's agent John Wyatt supplied Parker's Flint glass to Lunar Society members and associates (Birmingham City Archives)

The correspondence between Small and Watt makes fascinating reading and Small not only shows an interest in optics and steam engines but also on politics and the environment. In 1773 he expanded on his theory that the earth is cooling by suggesting that the 'frozen sphere was expanding by 300th part of a degree each year' which would result in the future with the earth being totally frozen like the Moon! He had it seems made measurements of the encroaching ice every winter while he was in America. Small also suggests that the constant explosions of black powder by the great powers was affecting the environment which could result in a perpetual summer – global warming!

Such free thinking stimulated Watt, Boulton and others into complex discussions.[48] Watt always kept himself up-to-date regarding developments in scientific instrument making, and as one can see from his correspondence he carried out experimental work improving a method of measuring the angular distance between stars, and developing a prismatic micrometer for which he never attempted to obtain a patent.[49] This micrometer would be the cause of some controversy.

In 1777 two papers appeared in *Philosophical Transactions* of the Royal Society regarding the improvement in micrometers. The papers were by Roger Joseph Boscovich (1711-1787) a Jesuit priest working at the University of Padua who referred to R. Fontana's and Alexis-Marie Rochon's (1741-1817) micrometers, with his design being an improvement on them.[50] The second paper was by Nevil Maskelyne (1732-1811) the Astronomer Royal who described in some detail his design for a prismatic micrometer stating that he was the true inventor and appended to the paper letters from Alexander Aubert (1730-1805) and Peter Dollond supporting his claim.[51]

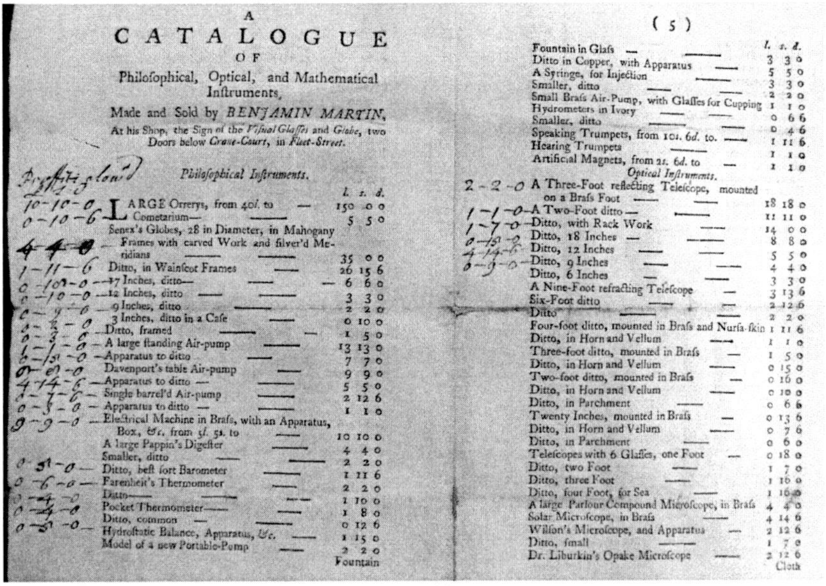
A
CATALOGUE
OF
Philofophical, Optical, and Mathematical Inftruments,
Made and Sold by *BENJAMIN MARTIN*,
At his Shop, the Sign of the *Vifual Glaffes* and *Globe*, two Doors below *Crane-Court*, in *Fleet-Street*.

Philofophical Inftruments.

	l.	s.	d.
LARGE Orrerys, from 40*l.* to	150	0	0
Cometarium	5	5	0
Senex's Globes, 28 in Diameter, in Mahogany Frames with carved Work and filver'd Meridians	35	0	0
Ditto, in Wainfcot Frames	26	15	6
17 Inches, ditto	6	6	0
12 Inches, ditto	3	3	0
9 Inches, ditto	2	2	0
3 Inches, ditto in a Cafe	0	10	0
Ditto, framed	1	5	0
A large ftanding Air-pump	13	13	0
Apparatus to ditto	7	7	0
Davenport's table Air-pump	9	9	0
Apparatus to ditto	5	5	0
Single barrel'd Air-pump	2	12	6
Apparatus to ditto	1	1	0
Electrical Machine in Brafs, with an Apparatus, Box, &c. from 5*l.* 5*s.* to	10	10	0
A large Pappin's Digefter	4	4	0
Smaller, ditto	2	2	0
Ditto, beft fort Barometer	1	11	6
Farenheit's Thermometer	2	2	0
Ditto	1	10	0
Pocket Thermometer	1	8	0
Ditto, common	0	12	6
Hydroftatic Balance, Apparatus, &c.	1	15	0
Model of a new Portable-Pump	2	2	0

Fountain

(5)

	l.	s.	d.
Fountain in Glafs	3	3	0
Ditto in Copper, with Apparatus	5	5	0
A Syringe, for Injection	3	3	0
Smaller, ditto	2	2	0
Small Brafs Air-Pump, with Glaffes for Cupping	1	1	0
Hydrometers in Ivory	0	6	6
Smaller, ditto	0	4	6
Speaking Trumpets, from 10*s.* 6*d.* to.	1	11	6
Hearing Trumpets	1	1	0
Artificial Magnets, from 2*s.* 6*d.* to	1	1	0
Optical Inftruments.			
A Three-Foot reflecting Telefcope, mounted on a Brafs Foot	18	18	0
A Two-Foot ditto	11	11	0
Ditto, with Rack Work	14	0	0
Ditto, 18 Inches	8	8	0
Ditto, 12 Inches	5	5	0
Ditto, 9 Inches	4	4	0
Ditto, 6 Inches	3	3	0
A Nine-Foot refracting Telefcope	3	13	6
Six-Foot ditto	2	12	6
Ditto	2	2	0
Four-foot ditto, mounted in Brafs and Nurfa-fkin	1	11	6
Ditto, in Horn and Vellum	1	1	0
Three-foot ditto, mounted in Brafs	1	5	0
Ditto, in Horn and Vellum	0	15	0
Two-foot ditto, mounted in Brafs	0	16	0
Ditto, in Horn and Vellum	0	10	0
Ditto, in Parchment	0	6	6
Twenty Inches, mounted in Brafs	0	13	6
Ditto, in Horn and Vellum	0	7	6
Ditto, in Parchment	0	6	0
Telefcopes with 6 Glaffes, one Foot	0	18	0
Ditto, two Foot	1	7	0
Ditto, three Foot	1	16	0
Ditto, four Foot, for Sea	1	16	0
A large Parlour Compound Microfcope, in Brafs	4	4	0
Solar Microfcope, in Brafs	4	14	6
Wilfon's Microfcope, and Apparatus	2	12	6
Ditto, fmall	1	7	0
Dr. Liburkin's Opake Microfcope	2	12	6

Cloth

Fig. 13 Pages from a catalogue of Benjamin Martin with James Watt's annotations (Birmingham City Archives).

This did not go unnoticed in Scotland. Two letters dated July 2nd 1778 were sent to Watt (then living in Birmingham) from Glasgow. The first was from Dr. William Irvine (1743-1787) a lecturer in Chemistry at Glasgow College where he writes:

> "Pray have you seen the last volume of Transactions (volume 67) you must surely know that it contains the description of a certain micrometer that shall remain nameless made by one J. Watt six or eight years ago & which has been in MacFarlane's Observatory in Glasgow for several years past. Would you not think the said J. Watt should claim this discovery?"[52]

Irivine adds that although Maskelyne has provided witnesses to his invention there were some in Glasgow who were willing to act as witnesses for Watt, namely Dr. Thomas Reid (1710-1796), Dr. (Alexander) Wilson, Pat(rick) Wilson (1743-1811), and Gilbert Hamilton (d.1808). Irvine was a good friend of Watt having studied under Joseph Black and had assisted him in his latent heat experiments. The second letter of the same date came from Patrick Wilson of Glasgow Observatory, the son of Alexander. At this time he was acting as his father's assistant (he would take over from his father in 1782). He no doubt used Watt's prismatic micrometer in his work; he states that he and his father feel that Watt's name should be put forward and that they would support him by providing evidence that Watt had a prior claim to the invention. [53]

A third letter arrived dated 3rd July from Gilbert Hamilton giving his support and again urging Watt to state his claim to the invention.[54] It seems from the correspondence from Glasgow University that Maskelyne was not a popular figure. He had visited the observatory in 1775 and a report of this visit was sent to Watt by Irvine in which he states that he and others questioned Maskelyne regarding achromatic lenses and Ramsden's ideas on eyepiece construction. His comments on Maskelyne are far from complimentary:

> "When Maskelyne was here he seemed to have no general opinion either of Ramsden or of his eyepiece but prejudice is no very rare plant in the soul of London."[55]

Given Maskelyne's visit to Glasgow in 1775 and the apparent poor impression he made I wonder if Irvine and Patrick Wilson were thinking that Maskelyne had copied Watt's micrometer. Interestingly Maskelyne's visit coincided with a robbery at the observatory, however the culprit was soon caught and all the stolen articles were returned, although Wilson adds an interesting comment:

> "The dog is taken who robbed our observatory and every article recovered – and a hand of suspicion surrounds as to certain folks."[56]

Watt it seems never tried to challenge Maskelyne, although Watt was always reluctant to challenge others on the priority of inventions or patents even though in several cases he probably had the right. This did not stop his friends often corresponding in support of him. It is interesting that soon after Maskelyne's paper, Jesse Ramsden submitted one to the Royal Society speaking of his improvements in micrometers.[57] Years later Watt's take on the invention was published by James Patrick Muirhead (1813–1898) in his three volume work on Watt's inventions. In this Watt describes his prismatic micrometer and states that it was later that Maskelyne and Rochon announced theirs, Watt also states that his micrometer was well known as he told people like Ramsden about it.[58]

Watt was casually acquainted with Boulton prior to the days of the steam engine business with their first meeting taking place in 1768 at Soho in Staffordshire. William Small had tried for some time to get Boulton interested in Watt's steam engine experiments, but Boulton had resisted getting involved.

Discussions on optics and lens making became a major part of the activities of the gathering of friends, which included at this stage James Keir and Erasmus Darwin. The name Darwin is mostly associated with Charles Darwin (1809-1882), Erasmus' grandson, yet it is Erasmus who is by far the more interesting and eclectic of the Darwin family. Although a physician his interests ranged wide including electricity, education, chemistry, botany, engineering, poetry, canals, geology, aviation and of course astronomy.

James Keir came from Muirton near Edinburgh and following the untimely death of his father was brought up by his mother with the help of George Lind, Lord Provost of Edinburgh. He spent eleven years in the army before resigning his commission at the rank of Captain to pursue his interest in chemistry coming into contact with Boulton via John Whitehurst in Derby. Keir was fascinated by the properties of glass and conducted many experiments to remedy the imperfections in flint glass, becoming a glass manufacturer setting up a works in Stourbridge. He also set up a chemical works in Tipton producing white and red lead for potters and glassmakers, and alkalis for soap. His expertise in glass would have been invaluable to Small and Watt in their experiments with optics. As a glass merchant Keir shipped glass around the country and through William Small he was put in touch with James Watt. In December 1771 Keir wrote to Watt to ask him about the flint glass trade in Scotland and in particular Glasgow, wishing to know the prices, discounts, terms and conditions etc, but also if Glasgow merchants were exporting to America and if Keir could take a share in this trade. Watt duly responded and supplied Keir with the relevant information and even arranged for his friend Gilbert Hamilton to take a consignment of glass from Keir.

Keir's own interest in optics can be gleamed from his extensive article on achromatic telescopes in his *'The First Part of a Dictionary of Chemistry'* published in Birmingham in 1789. It was the need to improve the quality of lenses and to obtain brighter images that became the Lunaticks' focus of interest. William Small was constantly experimenting stating to Watt:

> "...have many difficulties that lessen my confidence in observations made with the most accurately divided instruments. For example, no optical Instrument hitherto constructed Catoptric, or Dioptric, or Catadioptric produceth an exact copy on any object..... the unsteady refraction of light passing thro the atmosphere are also vile things; not those mentioned by Astronomers only, but others I will shew you when we meet..."[59]

Reflecting telescopes although able to gather much light still had to be made accurately with their mirrors being figured and polished. Glass mirrors were produced but were prone to cracking and were expensive to purchase and there was great difficulty in grinding large examples, mirrors of a metal alloy - speculum had their own problems especially in getting the alloy mix right (Boulton himself began experimenting with speculum as a material for mirror making noting a new procedure in 1778).[60] Erasmus Darwin came up with a novel idea, by using several lenses or mirrors light could be focused onto a common point. Darwin experimented using Dollond lenses and candles and found that trying to align all of the elements was a real mechanical challenge. He wrote in his commonplace book:

"Suppose 20 glasses, either lenses or specula are so placed as to throw all their image of a certain object onto one focus – there will then be one image with 20 times the brightness that one lens of speculum would produce."[61]

Fig. 14 Erasmus Darwin one of the greatest thinkers in the 18th century who foresaw the advent of the multi-mirror telescope, described a black hole in prose and imagined an ever expanding universe. (From a print in the Lound Collection)

This multi-lens and multi-mirror concept predates the Hopkins Observatory by over 200 years. Darwin was unable to make the multi-mirror or multi-lens system work noting:

"I try'd this with two foci from a candle and two of Dollond's acromatic object glasses, & they seemed to coincide, but not with accuracy."[62]

Although achromatic lenses were an improvement all depended on the quality of glass, Boulton had encouraged the use of flint glass obtained through William Parker but flint glass had imperfections, streaks and veins were present which greatly affected scientific instruments. James Keir in his Dictionary of Chemistry suggested the reason for this was that flint glass is composed of "materials of more different densities".[63] Josiah Wedgwood decided it was a problem he would try to solve.

Wedgwood is known as a Staffordshire potter, in fact he was from a family of potters but it was Josiah who would make the name famous throughout the world. He was an amateur scientist and he applied his knowledge to his industry developing methods for kilns, glazes, temperature measurement etc. His interest in glass came from necessity; he used a frit of ground flint glass in his ceramic process, he consulted Keir on this and the problems of heating vitreous substances. Keir told Wedgwood how to use raw flint glass and lectured him on annealing. According to Robert Schofield it was for this assistance that Wedgwood looked at solving the vein problem. He made notes in his *Common Place Book* referring to conversations he had with Keir on the subject. In March and April 1783 Wedgwood recorded the results of his experiments to *'Remedy the Imperfections of Flint Glass for Achromatic Instruments;'* experiments that took place at his own laboratory at Etruria, Mr. Knight's Glasshouse in Liverpool, and Mr. Holmes' Glasshouse in London. Wedgwood was able to replicate the imperfections taking detailed measurements using beams of light passing through various samples; thus coming to a conclusion as to the cause and a remedy.

Fig. 15 Josiah Wedgwood, potter and amateur scientist who found a solution to waviness in flint glass. (From a print in the Lound Collection)

Wedgwood's work on the flint glass problem is astounding starting in 1779 involving detailed experimentation at various sites across the country resulting in 1783 with his detailed paper entitled *'An Attempt to Discover the causes of cords and waviness in Flint Glass and the most probable means of removing them'*. Wedgwood offers a solution to the problem by constantly stirring the molten glass to maintain homogeneity or to remove samples from different levels of the melting-pot using each for different applications.[64] Yet after all this time and effort his paper was never published or presented to the Royal Society! Why? The result of his work was communicated to Keir, Ramsden etc but not through the formal channels of the Royal Society. The reason may well be cost; excise duty on glass was high – even if the glass was not used to make a specific item. It appears that one set of Wedgwood's experiments had to be cancelled due to a dispute between Mr. Knight and the Excise Officer. The publication of Wedgwood's work would have indicated how much glass he used and thus he may have had to pay a large sum in duty. Many writers on the history of glass have suggested that the excise duty on glass prevented experimentation and research by English glass manufacturers. Even so Wedgwood had solved a problem that had dogged opticians and this information was released, yet the pure cost in excise duty to produce the quality of glass using his technique was restrictive. Wedgwood's work predates that of Pierre Louis Guinand (1748-1824) by 15 years, which was repeated by Michael Faraday in 1830 as part of a Royal Society study to improve the quality of glass.[64]

Fig. 16 James Keir an expert in flint glass and chemistry
(Birmingham City Archives)

3. The Soho Observatory

Boulton, no doubt encouraged by his friends was becoming more absorbed in astronomy and once again James Ferguson became an influence as in 1771 he was back in Birmingham presenting more of his lectures; with the first of a new series starting on 22nd May once again at the Assembly Room in The Square. Ferguson stayed at Soho House as Boulton's guest for some of the time during his second stay in Birmingham;[1] although for most of the time he appears to have stayed with Richard Blockley, a locksmith and engine turner at 65 Bull Street.[2]

*Fig. 17 North West face of The Square. Ferguson's series of lectures in 1761 & 1771
were presented in the Assembly Room at number 11 (arrowed)
(From 'Memorials of the Old Square' by Hill and Dent)*

Ferguson's updated lecture programme was still divided into twelve lectures and the content would have had a particular appeal to Boulton and his friends as can be seen from Ferguson's published *Syllabus*:

1. Mechanicks demonstrated by the Lever, Wheel and Axle, inclined Plane, the Wedge, the Screw, and Models of two Cranes.
2. Wheel carriage, loading of wagons, and the best method of constructing Mills for grinding Corn and sawing Timber.
3. Hydrostaticks in which the laws of fluids will be explained.
4. Hydraulicks In which the different methods of raising water above the level of Rivers or Springs, will be shewn by a Variety of working Models of Pumps and other engines particularly the Hungarian Engine.
5. Nature and properties of Air which will be demonstrated by a Variety of capital Experiments on the Air-Pump.
6. Second Lecture on the Air-Pump.
7. Electricity in which besides the Experiments that are commonly shewn, he will set Models of Mills, Clocks and Orreries in Motion, by Streams of the Electrick Fluid; shew an Electrical Fire-Screw, a natural Representation of the Aurora Borealis, the Method of preserving Buildings from Damage by Lighting, and conclude with an Account of Medical Electricity.

8. The Laws by which the Planets move and are retained in their orbits, which
    will be confirmed by a great Variety of Experiments on the WHIRLING
    TABLE. The earth's motion round the Sun in a Year, and around its axis in 24
    hours, will be demonstrated; the Doctrine of the Tides explained at large, and
    the Cause of their rising equally high at the same Time on opposite Sides of
    the Earth, made evident to sight.
9. The Orrery and Cometarium in which the Motions and Distances of the
    Planets and Comets will be shewn.
10. The Orrery in which the Causes of the different Lengths of the days and
    Nights, and consequently of all the Vicissitudes of Seasons depending
    thereon, will be clearly explained and demonstrated; so will the amazing
    Velocity of Light be, and the Method of finding the Longitude by the Eclipses
    of Jupiter's Satellites.
11. Harvest Moon and the Phases of the Moon.
12. The Orrery in which the Causes, Times and Returns of all Eclipses of the Sun
    and Moon will be demonstratively shewn.[3]

Once again the lecture series was a tremendous success with no doubt many of
the Lunaticks attending as well as many other locals. A second series began on
10th June[4] and again demand was high so a third series was announced to begin
on 1st July and then a fourth series was advertised on 15th July! Ferguson enjoyed
meeting with people on his tours and from his various letters he seems to have
had a low boredom threshold. It was not uncommon for him to become involved
with local people. It was reported in *Cornish's Stranger's Guide To Birmingham*
(earliest edition 1839) that 'Ferguson the astronomer' placed a Meridian Line on
the South side of the tower of St. Martin's Church in the Bull Ring. This is the
kind of thing Ferguson might have done. Cornish's book does not provide a date
when this event occurred but it would have been either 1761 or 1771. The church
was in poor repair in 1753 and in fact the tower had been struck by lightning three
times – something that would have interested Ferguson especially as his seventh
lecture refers to lightning conductors. The tower was eventually repaired in 1781
and what might have appeared to be a meridian line could have been a result of
the repair work. As Ferguson died in 1776 he could not have been involved
during the repairs. No other text on the history of Birmingham including a
contemporary text by William Hutton mentions a meridian line on the church.[5]

During his visit Ferguson and Boulton no doubt discussed various matters[6]
including the designs for two sidereal clocks then in the planning stage; and of
course optics and the use of telescopes as a letter from Ferguson to Boulton
written from Kidderminster in August 1771 demonstrates:

"The number of my sufferers is only 26 – I shall finish next Thursday and be glad
to get out of this place, where I have not had six hours of tolerable health... I am
happily lodged with one Mr. Howell a Barber in this place, who is really very
versible, civil, and scientific. He understands optics, Electricity, and astronomy
very well; has been about constructing an orrery, has a very good microscope and
a tolerably good Reflecting Telescope of which he occasionally makes a
microscope in a way I never heard of before, which is this. He has cut out a square
hole in the Great tube of the Telescope even with the Focus of the little speculum
and in that place he puts his microscopial object (softening the tongs or glass that

hold them by a screw to the edge of the hole) and by looking in the usual way through the eye-glass Tube the object is greatly magnified, and the vision is very distinct. When he is done he shuts the hole in the tube by a little hole that turns on hinges – Pity this man was made a poor Scull-thatcher for he might have shone in a much higher sphere of life..." [7]

The 'scull-thatcher' is Thomas Howell (d.1793) who lived in a house in the Bull Ring area of Kidderminster. He was considered by many in Kidderminster to be a gentleman of superior character.[8] His telescope-microscope sounds a fascinating instrument, the tube was cast in pewter by Howell himself with the optics being obtained in Birmingham; one can only lament at not having it available for examination.[9] Howell owned an orrery and Ferguson offered advice on how to incorporate a mechanism for reproducing the moon's phases even speaking with a local clockmaker.[10] Of great importance is the fact that Ferguson is clearly presenting a series of lectures in Kidderminster (to only 26 people). Although it was known that Ferguson went to Kidderminster following his lecture series in Birmingham it was not known whether or not he was lecturing. The late John Millburn who wrote papers and an excellent biography of Ferguson commented:

"After concluding his fourth course at Birmingham at the end of July (1771), Ferguson apparently went to Kidderminster. However as the town is only 27 km from Birmingham he may have just been passing through."[11]

The discovery of Ferguson's letter to Boulton in Birmingham Archives confirms that he was indeed lecturing in Kidderminster possibly at the invitation of local non conformists including James Hill, Thomas Howell (d.1793), Job Humpage and Dr. James Johnstone (1730-1851) beginning the lecture series on 31st July and ending on 15th August 1771.[12] One attendee at three of the twelve lectures in Kidderminster was James Hill's son Thomas Wright Hill (1763-1851) who many years later would be associated with Joseph Priestley who encouraged him into becoming a teacher; which he did, opening a school (Hazelwood on the Hagley Road, Edgbaston) where astronomy was included on the syllabus.[13]

After Kidderminster Ferguson moved on to complete another two series' in Worcester,[14] before moving on to Newcastle Under Lyme where he dined with Josiah Wedgwood. Wedgwood was away on business when Ferguson began his lectures and missed the first two, which disappointed him as he wrote to Thomas Bentley:

"...got home the next day to dinner & found all well there – Except that Mr. Ferguson had given two of his Lectures at Newcastle upon Mechanicks which was not so well for I should have been glad to have seen them, but to make amends he is to dine with me to morrow when I am to give him some lectures upon Pottmaking & hope to receive some of his good things in return."[15]

It was around the time of Ferguson's visit that Boulton began to consider the building of a room to house his growing collection of books and scientific instruments as well as a place where he could study his various interests. He had moved into Soho House in 1766 a farm house situated in the countryside just outside Birmingham, and from the start he considered modifications to the building. He wrote in his notebook:

> "A round room for my study, library, museum or hobby horsory to hold 6 handsome bookcases with drawers in the lower parts to hold things which relate to subjects of the books which (illegible) are in upper parts E:g: A book case containing chymical books should have drawers under which contains metals, minerals, and fossells - between each bookcase should be a sophie and under the space between ye upper parts of ye cases should be fixed such instruments as barometer, thermometer, pyrometer, quadrants, all sorts of optical, mathematical, mechanical, pneumatical, & philosophical instruments, also clocks of sundry kinds both geographical, syderial, lunar, solor system and one good regulator of time. A table in the middle of ye room & a skylight in ye middle of the domical roof which roof may be covered with sail cloth or brown paper. Out of this round room should open a private door into a passage in which passage should open doors into sundry conven(ien)t rooms such as cold and warm bath a laboratory a dressing and powdering room and an observatory for a transit instrument etc." [16]

The domical roof he refers to suggests he would like to have an observatory as part of his house, he does however also consider a separate observatory building. In a very important entry in his notebook he states clearly what type of building he wants.

"OBSERVATORY

> A round building with an iron toothed wheel upon the wall plate to turn the top round the top may be covered with brown paper or tin foyl."[17]

His plans are more involved than a simple little facility, at this period he hoped to build the observatory on the top of the house,[18] otherwise he will have a separate building; however, while plans are being made for the main observatory he also looks at a temporary building that he could use in the mean time.[19] It is an exciting (but perhaps not too unexpected) that Boulton looks at rotating the observatory roof, with the construction using brown paper or 'tin foyle' to enable a lightweight structure very suitable for a rotating roof. The instrumentation he plans to install is also listed along with some notes.

> "A transit
> A Quadrant upon a stand
> A good telliscope (sic) for occultations
> A register for rain
> Do Barometer
> Do Thermometer
> Do electricity
> Do evaporation
>
> If I could not have a circular top at the top of my house.
> What are all the things done by astronomers.
> An equal altitude Instrument
> The perpendicular
> A temporary observatory" [20]

He refers to the fact that he would like a circular top at the top of his house and I wonder if he is thinking initially, of a design similar to that of George III's Observatory at Richmond. He has also noted from whom he intends to purchase the equipment – Jesse Ramsden, seen by many as the greatest scientific instrument maker of the eighteenth century. Boulton lists the Ramsden order:

"3 thermometers
A Camera
A Hadley's Quadrant
A good perpendicular barometer
A pulveometer

To fit observatory
A Dividing machine
Clocks to follow the stars by
Dividing machines.

Plus:

Concave Mirror
Convex Mirror
Micrometer for my telescope.
I could not make my solar microscope into a magic lantern.
Prism with movable sides for trying of refraction and disipation of fluid.
The best level for water courses
Asimuth compass
Astronomical (illegible)
Botanical thermometer" [21]

He had already obtained a number of instruments which included 3 microscopes and a four foot telescope.[22]

Astronomical observatories in the 18th century were often octagonal towers designed to enable telescopes and quadrants to be placed on a flat roof or telescopes to be placed through slits in windows or through a slit on the south facing section of a roof. Towers, unless built extremely well had a stability problem which led to the development of buildings with several storeys.[23] Matthew Boulton wanted to overcome this by having the observatory on the roof of Soho House or a large extension added to it. Prior to this being constructed he wanted a small building for his astronomical activities thus a tower was not practicable especially if it was to be superseded in a few years. Notebook 8 has many pages that are unused, as Matthew Boulton's use of notebooks was haphazard, jotting down notes as and when he needed to and not always on sequential pages. Thus it was with delight that while flipping through the empty section of the notebook an illustration was found, Matthew Boulton's sketch of his proposed observatory.

The illustration (Fig. 19) shows the central round room with the domical roof, on either side are additional rooms with slits where a transit instrument would be fitted.[24] One can also see the skylights one above the entrance door and the other

on the dome wall. The equipment he listed would fit well into such a building; he
had clearly given a lot of thought to the project.

Boulton is absorbed with optical equipment and has begun his own work on
instruments. Boulton produces quality tube for the Scottish Hydrographer
Alexander Dalrymple's (1737-1808) surveying equipment.[25] Dalrymple sent a
glass to Boulton to have inserted some cross wires [26] as well as other instruments.

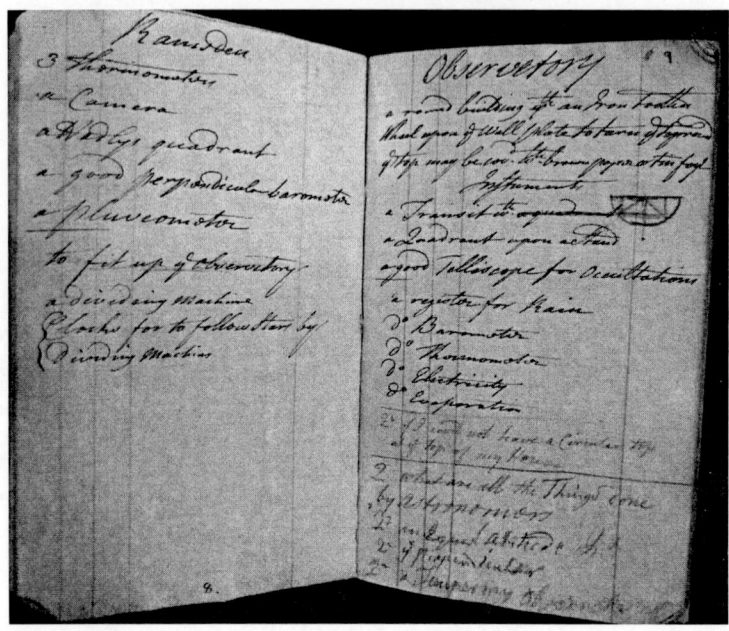

*Fig. 18 Pages from one of Matthew Boulton's Notebooks referring to the observatory.
(Birmingham City Archives)*

In April 1775 Dalrymple was going to India on behalf of the East India Company
to serve as a member of the Madras Council, he would be travelling with the
Governor of Madras, George Pigot (1719–1777). His equipment which included
equatorial instruments was with Boulton who made plans to send them to
Dalrymple via Pigot. In a letter to Watt, Boulton comments that at the same time
he will send his telescope to his agent William Matthews (in London) in order to
arrange for a simple equatorial mount to be made for it. [27] This telescope is a 42
inch long mahogany tube achromatic refractor with a 3¾ inch object glass.

Boulton was not the only one buying astronomical equipment; Watt purchased
a telescope from Jesse Ramsden with which he was very happy. This telescope
was 2½ inch in diameter, with a 25 inch long tube and a 9 inch long screw-in
eyepiece.[28] Watt seems to have been searching for quite a while for a decent
telescope and typical of Watt he wanted to try out various types. Boulton wrote to
him while he was in London in 1775:

"...I hope you get a peek at the Moon or at Saturn through Short's telescope..."[29]

Fig. 19 Matthew Boulton's proposed astronomical observatory 1772, this building's purpose seems obvious to an astronomer when compared to Boulton's own description, but the image was thought to show a summer house by previous researchers. (Birmingham City Archives)

Watt had already written to Boulton saying that he had seen the telescope but had not looked through it due to bad weather, however he described it:

> "It is a Gregorian telescope with a similar one of the same kind for an eyepiece he (Short) thinks nothing of amagnifying power of 44 thousand times and says they are sufficiently distinct & light when used celestialy – he says the atmosphere is not magnified except by the first telescope, this I do not comprehend."[30]

Watt doesn't seem too convinced about Short's comments, magnifying powers of the period were not literally as high as stated, there was a great deal of exaggeration and 44,000 is a ridiculously high figure.

The quality of the telescope supplied to Watt was so good that Boulton wanted a similar one although of larger aperture produced by Ramsden. Ramsden wrote to Boulton on 24th August 1775 commenting of his distress that the object glass he is preparing for Boulton's somewhat larger telescope is of too poor a quality for the use intended. [31] However he does explain that a good equatorial mount is ready and can be shipped to him – this is the mount for the telescope Boulton sent to London with Dalrymple's equipment. A box of instruments (presumably for the observatory given the date) is also ready for shipping and Ramsden is asking for some money, "..favor me with 30 or 40 £ .."[32]

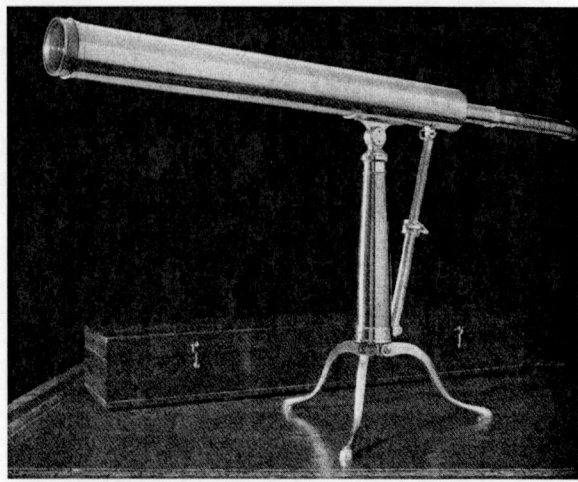

Fig. 20 Telescope made by Ramsden for James Watt that was sold at auction in 2003 for over £9000. (Courtesy Sotheby's)

With Boulton's enthusiasm fired up did he in fact build the observatory? The first record that an observatory actually stood in the grounds of Soho House comes in a letter of 1778 from the astronomer Alexander Aubert to Boulton and Fothergill from whom he had purchased a picture of 'Penelope':

"Gentlemen

I have received the picture of Penelope and I like it much –
When I was at Soho I had sight of a little round building in which I observed a telescope of large aperture and could but lament it having suffered much by the rain & weather which the roof did not shelter it from. I happen to be in immediate want of such a telescope consequently I should like to know if Mr. Boulton will part with it; I suppose the object glass is unhurt and I desire of Mr. Boulton no more than such equitable terms as the general stake of the instrument will deserve.

I am with great regards,
Gentlemen, Your most humble and most
Obedient servant
Alexr Aubert.

Pray who was the telescope made by
And is it or was it a good one." [33]

Aubert had an extremely well equipped observatory at Loampit-Hill near Deptford (later moved to Highbury). From his description the building's roof seems to have suffered and with it being made of brown paper or foyle as Boulton indicated then without regular maintenance it would have soon been damaged. Boulton had a number of outbuildings constructed in the grounds of Soho House and identifying a small round building should be easy. However records of buildings from this period are sketchy. Two round buildings appear on hand drawn plans. One is a round boiler house the other is the famous Hermitage - a wooden building with gothic style windows and a thatched roof located in a secluded spot amongst some trees. Boulton used this small hide-away as a place of contemplation.

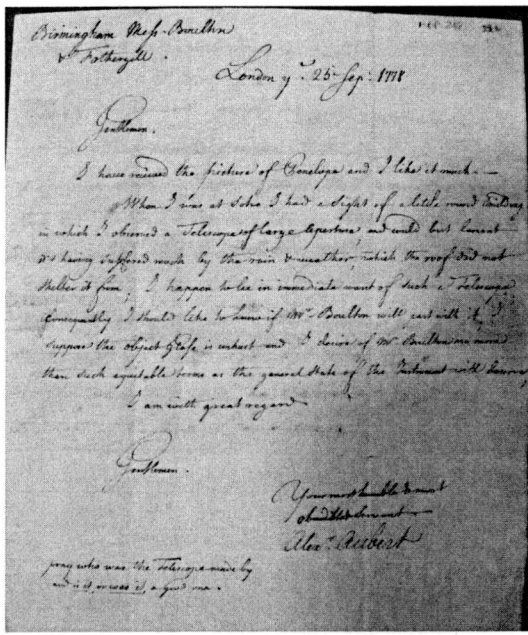

Fig. 21 Alexander Aubert's letter to Matthew Boulton asking after a telescope which he saw in a little round building. (Birmingham City Archives)

An album of drawings came into the possession of Birmingham Museum and Art Gallery that includes a third round building. The album of drawings was made between 1796 and 1798 by John Phillp (1780-1820)[34] a friend of Matthew Boulton who worked as a designer at the Soho Manufactory who for a time lived at Soho House. While in residence he made a number of drawings of the buildings on the estate.

The building illustrated (Fig.22) was small and round as Aubert had intimated. A single-story octagon, twelve feet across and stood just over ten feet tall. Only one face of the building showing an entrance door and two windows is drawn. The windows are of gothic style that is in keeping with the style of windows of other outbuildings on the Soho site. The octagonal roof sits on a decorated support, from the illustration it is not possible to say whether or not the roof rotated as per Boulton's suggestion although the function of the building is likely to have changed since 1778 especially as Boulton indicated a domical roof and his original drawing showed such a feature, so if the structure of this building was the building seen by Aubert it is likely the roof had been replaced by 1796. Unfortunately no other illustrations of the building have been found to date. It is likely that had Boulton's original design been built Aubert would have referred to it in more detail than *'A little round building'*.

*Fig. 22 Possibly Boulton's Observatory building at Soho in 1796 From John Phillp's
Album. Is this the structure of the little round building seen by Aubert?
(Birmingham Museum & Art Gallery)*

This building and Boulton's original observatory design are interesting; they
are early examples of a single story observatory building. Of great interest is a
similar design used for the Copenhagen Observatory started in 1778 and
completed in 1780 (Fig.23) which replaced a tower observatory.[35] It has been
speculated that the Copenhagen Observatory was inspired by Carl Hårleman's
Observatory in Stockholm that was completed in 1753.[36] This was a domestic
mansion with an observatory forming the lower part of the building with long
windows and a residence on a third floor. Dimensionally the octagonal section of
the Copenhagen Observatory is twice the size of Boulton's octagonal building
and has the two additional transit wings as seen in Boulton's original design, the
similarity is striking. What is more tantalizing is that Thomas Bugge (1740-1815)
was appointed director of the Copenhagen Observatory in 1777, his first task was
to consider rebuilding the facility since the old tower was in poor condition.
Bugge set out on a tour of Sweden, the Netherlands and Great Britain in order to
look at different astronomical observatories in order to decide on a design for
Copenhagen. He visited London, Cambridge and Oxford between September and
November 1777 making notes of observatories including Greenwich, Trinity
College, St. John's, and Radcliffe noting their contents in some detail. He also
spent some time at Richmond to examine the King's Observatory. Bugge also
visited instrument makers including Jesse Ramsden, Edward Nairne, Samuel
Witford and Benjamin Martin, and spent some time with Alexander Aubert at his
observatory in Deptford. Even though Bugge met with several people well known
to Boulton there is no evidence that the two men actually met.[37]

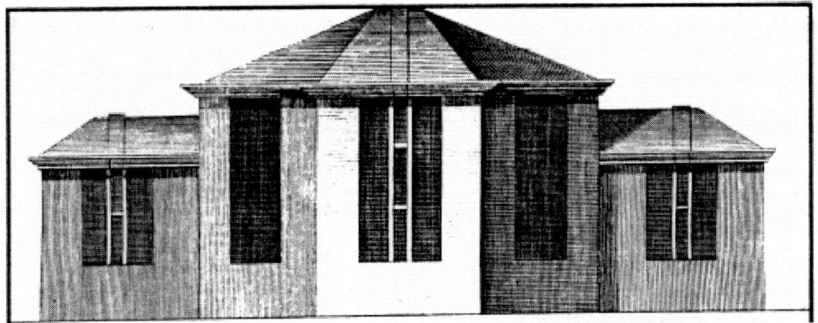

Fig. 23 Copenhagen Observatory 1778.
(Observationes Astronomique 1784 by Thomas Bugge)

The general look of the octagonal building with Gothic styling is one of several outbuildings at Soho which were designed by Boulton's favourite architect James Wyatt (1746-1813) who was well known for his work on the octagon tower Radcliffe Observatory.[38]

Figure 24 shows how the octagonal building would have looked with a domical roof on the top. Although this is the only small round building that has been discovered, there is a possibility that Boulton had started construction of his original design but only completed the central round portion which was then seen by Aubert. Unless new material is found, one can only speculate.

What was inside the observatory? Aubert in his letter refers to an *'equatorial of large aperture'* but no further details. The list of items in Boulton's notebook may give us a clue but without anything firm one can only guess. From snippets of information from William Small, James Watt, Alexander Aubert, and Boulton's correspondence with Ramsden one can see that Boulton owned a large quantity of astronomical, meteorological and other scientific instruments. Unfortunately personal letter books covering the years 1768 -1775 have not been found and examination of the letters of that period that do exist one can see gaps in the material.

Fig. 24 How the small round building might have originally looked with a domical roof (Artwork by author)

According to Aubert by September 1778 Boulton's observatory roof was in a poor state of repair. In a letter to James Watt, Boulton states that he observed the partial eclipse of the Sun of 24th June and showed it to Watt's daughter Peggy (Margaret) (1767-1799), the roof and the telescope were probably fine then.[39] Yet it does seem that Boulton had reduced his astronomical activities to merely making the odd technical note on telescopes and optics in letters but there is no record in his notebooks of any further developments at the observatory.

The other members of the Lunar Society too were concentrating on other matters especially after the death of William Small who did appear to be a major force behind the optical research and was a firm favourite among all of the Lunaticks, although astronomical instrumentation was still of interest.

What had happened to force Boulton away from one of his two favourite pastimes?

In 1774 just a short period after he had the observatory built he began working with James Watt on the development of steam engines. James Watt's financier John Roebuck (1718-1794) had gone bankrupt and for some years had been struggling to keep James Watt's steam engine experiments going. Boulton was in need of a steam pump to supply water from Hockley Brook to the Soho Manufactory. Watt had always been enthusiastic about bringing the steam engine to Soho where he felt the workforce was better skilled and able to perfect the engine. Thus in lieu of money owed to him by Roebuck he agreed to support James Watt's improved steam engine eventually forming a partnership with Watt for which the world knows the name Boulton. On May 31st 1774 James Watt arrived in Birmingham and Boulton's life changed as he began a new enterprise;[40] with his time taken with the new enterprise there is no wonder that the observatory was in disrepair. Boulton's interest in astronomy did not cease completely but was placed very much to the back of his mind as one can observe from his response to Aubert's enquiry about the large aperture telescope housed in the observatory:

> "Sorry that I'd not met him
> Telescope I'd not intend to sell but some other equatorial instruments would be more convenient.
> 2 nocturnal eyepieces & 1 eyepiece for daylight
> 1 eyepiece with a concave glass that magnifies about 200 times paid Mr. Dollond £20 or £21 - added an Equatorial apparatus which has not been yet (illegible) – Ramsden
> -   Dollond in good (illegible) of glass (illegible)
> -   (consid)er it as more valuable than when I purchased but would not sell it for less than £20. I have always considered it as an exceeding good one, the best of the kind I have seen" [41]

Aubert responded to Boulton's note accordingly:

> "I have been favoured this morning with your obliging letter and I return you many thanks for your readiness to dispose of the instrument on the consideration of my being in want of it. I have not the least doubts but that although in appearance hurt it is in substance a good one and if you will be kind enough to let it be packed carefully and sent to me as soon as possible, I will pay the £20 you

think I should give for it, to whoever you please the carriage will & ought to be at my charge. I beg the favour of you to observe before it is packed that the cell containing the object glass be secured quite home and I had rather the whole (I mean telescope, stande etc) should be sent in their present condition, than have anything done to them to make them look better – when you come to London I shall be very glad to have the pleasure of being personally acquainted with you, men of genius are scarce. I have an observatory very near Lewisham in Kent where I shall also be glad to see you" [42]

Although from the letter I wonder if Aubert is not at cross-purposes with Boulton as to which instrument he is buying! The equipment that Boulton sold to Aubert was a 42 inch long achromatic telescope with mahogany tube and 3¾ inch object glass by Dollond mounted on a mahogany block with equatorial motions and rack work by Ramsden. This is the telescope Boulton sent to London with Dalrymple's equipment to have an equatorial mount added. Following his death Aubert's entire astronomical collection was auctioned with the telescope he purchased from Boulton being referred to as *'The Bolton'* being sold for £53 11/-. [43]

Certainly Boulton was not expecting his time to be taken up by the engine business as much as it did as he continues to order astronomical equipment. In 1779 he ordered a 30 inch achromatic telescope from Nairne and Blunt.[44]

Fig. 25 Alexander Aubert, gifted amateur astronomer who saw a small round building containing a telescope of large aperture at Soho, the only eye witness account of Boulton's observatory. (From a print in the Lound Collection)

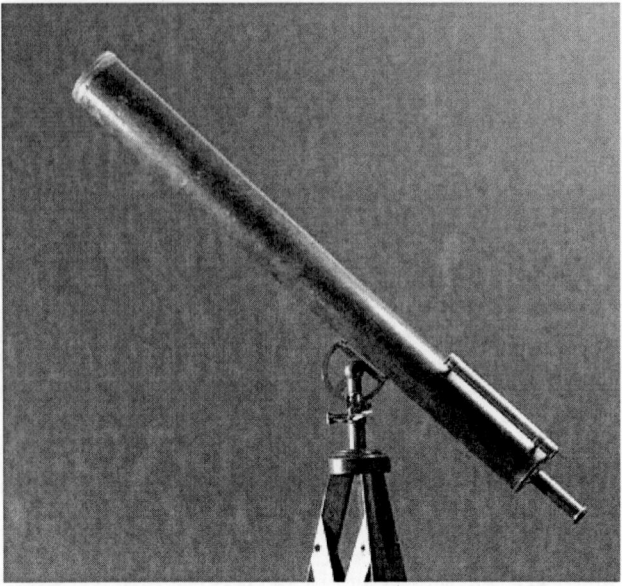

Fig. 26 A 42 inch long Achromatic Refractor with Mahogany Tube and 3¾ inch object glass by Dollond of the type Matthew Boulton sold to Alexander Aubert. The difference in the picture to Boulton's is the mount; Boulton's was an equatorial by Ramsden (Museum of the History of Science, Oxford)

Boulton's astronomical friend James Ferguson died in 1776 but he was not the only itinerant lecturer to become so closely involved with the Lunaticks. John Warltire (1725/6–1810) had begun lecturing in around 1763 and worked with Erasmus Darwin, Joseph Priestley and William Withering carrying out experiments - mainly on air.[45] It may be due to these connections that Warltire came to Birmingham in 1776 to present the first of many series of lectures he gave in the Town. Josiah Wedgwood employed him as a personal tutor to his son John in 1779. Warltire developed a friendship with all of the Lunar members, although he had given astronomy lectures his interest with the Lunaticks was mainly in connection with chemistry. The other lecturer that was well connected with the Lunaticks was Adam Walker (1730/31–1821). He was a friend of Ferguson, and purchased some of William Griffiss' equipment in 1766.[46] Walker though was one of *the* great scientific showmen of the 18th century combining music, props and wonderful performances to enthral audiences. His first performance in Birmingham was in 1781 presenting a series of six lectures in the Assembly Room at *The Square*.[47] A year later his spectacular astronomical presentation *'Eidouranion'* or *'Transparent Orrery'* was staged at the *New Theatre* in New Street - it needed a theatre as the globe of the earth in the performance was two feet in diameter! The by-line for the lecture was *'the Mysteries of Astronomy the Entertainment of an Evening'*.[48] There is little doubt that many of the Lunaticks attended and the fact that they invited Walker into the conversations demonstrates the sense of wonder they all had for the universe.

4. Fireballs and Firebrands

Glasgow University was probably dismayed at losing Watt to Birmingham. The MacFarlane Observatory had been set up using the equipment he had repaired and included his prismatic micrometer. It no doubt would have been expected that Watt would have been on hand to clean and repair instrumentation from time to time. So his move to Birmingham may have been something of a shock. Certainly in one letter he received the opening makes this clear:

"What in the universe are you doing so long in this place of buttons and japanned crokery called Birmingham?"[1]

The distance between Watt and Glasgow University was in mileage only, as communication was constant. Watt's views on optics, instrumentation and other matters were sought especially on the types of glass used for lenses. Ramsden had apparently stated that lenses ground in the common way could be used for achromatics whereas Dollond and Short disagreed. Watt's views were sought by William Irvine who hoped that Watt would pay some attention to construct some achromatics and to procure some flint glass that could be used in Glasgow. He ends the letter:

"...send us at least some fragments of the mantle which you so uncomfortably took along with you."[2]

When Patrick Wilson took over as professor of Astronomy at Glasgow University he kept up a correspondence with Watt regarding the latest developments:

"The new planet has been found out in Mayer's catalogue of 1756 and its place determined for that timing – it has been found as a small (illegible) star in Tycho's catalogue and gives its place as a very long period and ascertains is of a planetary nature...Mr. Herschel told me he has seen the planet with his naked eye..."[3]

The new planet referred to is *Uranus* discovered by the amateur astronomer William Herschel from Bath in 1781. Herschel was a music teacher who had left his native Germany to pursue a music career however he had a passion for astronomy building his own telescopes, casting the mirrors in his home in much the same way that Boulton had done several years earlier. He received encouragement in his hobby from several people including John Michell. Herschel was assisted by his sister Caroline who became a decent astronomer in her own right. The new planet was initially referred to as *Georgium Sidus* (George's Star) named for King George III who like Herschel was Hanovarian. The discovery (which Herschel at first thought was a comet) brought him great fame, a pension of £200 per year from the King and a move from Bath to be nearer to his Royal supporter which naturally brought him into contact with the most eminent and influential people in Britain.

The letter from Wilson is dated 22nd September 1783 and although Herschel's planet is a talking point the big story is a more dramatic event visually. On the evening of August 18th a large bright meteor streaked across the skies of Britain and France. Wilson is keen to know if Watt had seen it:

> "Have you had any good observations of the big meteor? It was luckily seen from the terrace at Windsor by (James) Lind, (Tiberius) Cavallo, Sandby and Lockman, and their account accompanied with a drawing by Sandby has been lately given to the President of the Royal Society. Some good observations have also come to hand from various parts of England, Scotland and North of Ireland." [4]

Watt's friend Gilbert Hamilton also wrote giving a vivid description of the meteor's appearance in Glasgow:

> "Our weather at present is very hot & there has been an uncommon quantity of electrical material in the air – on Monday night last a large ball of fire with a long train behind it passed near the Town & during its passage the streets were as bright as if in sunshine." [5]

The other reports included some from Alexander Aubert, Nathanial Pigott (1725-1804), William Cooper, General James Murray (1721-1794) and one from a Lunar member - Richard Lovell Edgeworth; he observed the meteor from Edgeworthstown near Mullingar, Ireland. He stated that he saw the meteor at half past nine in the evening, its size was about one third of the moon's diameter, with a luminous tail twenty or twenty five of its diameter in length, moving from the north with an elevation of about 10 – 12 degrees above the horizon. He saw it for about 15 seconds and noted:

> "It exhibited the most vivid colours; the foremost part being of the brightest blue, followed by different shades of red." [6]

Edgeworth's interests generally lay in education and mechanics designing a number of carts, and a portable railway submitting several papers to the Royal Society. The meteor report is his only recorded astronomical comment although his daughter Maria (1767-1849) had a particular interest in astronomy visiting the Herschels at Slough with William's son John showing her how to use their 20 foot telescope observing 'Saturn and his rings, and the moon and her volcanoes'. The volcanoes that Maria referred to was luminescence that Herschel first observed in 1783 which he interpreted as volcanic activity. She also entertained John Herschel when he visited the Edgeworths in Ireland. [7]

The meteor was one of the great events of the period as the origin of such phenomena was open to speculation; indeed the August meteor was not the only one that had appeared, which prompted Nevil Maskelyne to draw up a method to observe meteors scientifically, sending his plan to William Herschel adding that "no-one is more likely to see them than yourself and no-one is a more diligent observer of the heavens." [8]

The appearance of balls of fire with sparks resembled electrical activity and thus many felt that the occurrences were caused by electrical activity in the atmosphere. Electricity was though the subject that all professional and amateur scientists wanted to play with. In Charles Blagden's (1748-1820) report to the

Royal Society on the event he leans towards the electrical theory for the phenomena.[9]

Fig. 27 Richard Lovell Edgeworth, his observation of the 1783 fireball formed part of Charles Blagden's report to The Royal Society.(From a print in the Lound Collection)

Of course the man for electricity and natural phenomena was Erasmus Darwin. He saw the meteor from Radbourne Hall near Derby, he duly wrote to the Royal Society about his observation:

> "…observed the meteor pass from north-west to north-east, and give out numerous large sparks just before it was conceal'd by the cornice of the corner of the house. This part of the cornice I accurately attended to, and also to the height of my eye against the window-frame: and on the next morning found a line drawn from these two points lay, as nearly as could be easily measured, at an angle of forty-five degrees."[10]

Darwin's view on fireballs was that they were formed by electrical discharges at a height of between 60 and 70 miles above the earth.[11]

Lind's observation of the meteor from the terrace at Windsor highlights just how much of a focal point the royal circle was regarding science. Lind often received visitors but he complained to Boulton that Windsor was a "Damned gossiping place" which prevented him from writing more often.[12] Even so many of the visitors were very welcome such as Patrick Wilson who soon after he observed the meteor from Glasgow was at Windsor staying with Lind but spending his nights observing the stars with William Herschel. Patrick and Alexander Wilson were in constant written communication with Lind with the Wilson's giving details of their astronomical work and Lind letting them know what Herschel was up to.

Fig. 28 Painting of the 1783 fireball as seen from the Terrace at Windsor by Lind, Sandby, Cavello and Lockman, Painting is by Paul and Thomas Sandby. (From a postcard in the Lound Collection)

Lind promised to write to Boulton if "Anything strange happens to the Georgium Sidus or the moon."[13] Boulton and Watt were often at Windsor and although we today would see Herschel's standing as the great astronomer making him sought out by the great and the good, in the 1780s it was actually Boulton and Watt who were sought out and their visits to Windsor caused great excitement. On a flying visit they both made in 1787 to see the King - Lind, André De Luc and Herschel missed seeing them. De Luc's daughter wrote to James Watt saying:

> "Afterwards we have been despaired at not seeing you Sir nor Mr. Boulton at Windsor, you may guess easily if you have a just idea of our sincere regard and affection. Dr. Lind & Dr. Herschel have also partaken of our regrets they both desire their kind remembrance to you."[14]

Both men were highly respected in the scientific community, however it was a period when there seemed to be a glut of people who shared an interest in the sciences and communicated with others. The Lunar Society was the prime example of this.

As already stated William Small's death in 1775 ended much of the Lunar Society's experimentation in optics. He was essentially replaced in the group by Dr. William Withering a doctor of medicine having graduated from Edinburgh University joining the staff at Birmingham General Hospital in 1779. He is more noted for his work in Botany and the discovery of Digitalis for medicinal use but Withering also had an interest in astronomy and was keen to obtain a large telescope. Boulton was checking out telescope makers and suggested that "Simons an optician at the corner of Marylebone" might be able to provide Withering with the instrument he needs.[15] Withering eventually purchased an 18

inch reflecting telescope from James Simons in 1782 and housed it at Edgbaston
Hall in Birmingham (now the club house at Edgbaston Golf Club).[16]

*Fig. 29 William Withering the discoverer of digitalis who purchased a telescope from
Sisson. (Birmingham City Archives)*

One of the most well known Lunaticks is Joseph Priestley, he did not move to
Birmingham until 1780 although he had been in correspondence with members of
the Lunar Society and had numerous common connections such as John Michell
and Benjamin Franklin. Born in Leeds and raised as a non-conformist eventually
becoming a Unitarian Minister in Leeds and Birmingham. Priestley was a
prodigious learner studying at least seven languages, history, philosophy,
mathematics and science. His expertise in optics has been overshadowed by his
work in chemistry. It was Benjamin Franklin who encouraged him to publish a
work on electricity and it was this interest that drew him to Matthew Boulton.
Priestley's *'History of Electricity'* was published in 1767 and was a great success.
He followed this book with *'The History and Present State of Discoveries
relating to Vision, Light and Colours'* assisted by John Michell. This book
includes the history of optics including the development of the telescope but was
not so successful; the initial list of subscribers included Henry Cavendish, Josiah
Wedgwood, Charles Lincoln, John Smeaton and Benjamin Franklin. Although
commercially unsuccessful it remained the only English language book on the
history of optics available for many years. From then on he concentrated more on
chemistry and politics.

In 1771 Joseph Banks (1744-1820) invited Priestley to join him on his second
voyage to the Pacific Ocean with James Cook (1728-1779) as Astronomer. Banks
had equipped and travelled on Cook's ship *'Endeavour'* on its first Pacific voyage

in 1768-71 now he was looking to do the same. It was an offer that pleased
Priestley enormously as he wrote to William Eden:

> "I think myself greatly honoured by Mr. Bank's desire of my accompanying him
> in his second voyage, in so respectable a capacity as that of Astronomer; and, to
> be as frank and explicit with you as the friendship you have shewn me requires, I
> will at once acknowledge, that the proposal will not be disagreeable to me...."[17]

*Fig. 30 Joseph Priestley the scientist Unitarian whose work on optics was
recommended to William Herschel. (From a print in the Lound Collection)*

The initial excitement soon turned to disappointment as Joseph Banks
informed Priestley that the decision as to who the astronomer on the voyage
should be would be determined by different professors of Oxford and Cambridge
Universities and as these professors were all clergymen they would more than
likely have scruples regarding the head of religion. Banks felt that given
Priestley's non-conformist views he did not think he could nominate Priestley let
alone have him accepted! Priestley responded to Banks:

> "Now what I am, and what they are, with respect to religion, might easily have
> been known before the thing was proposed to me at all. Besides, I thought that this
> had been a business of philosophy and not of divinity. If however, this be the case,
> I shall hold the Board of Longitude in extreme contempt..."[18]

In *'The Memoirs of Dr. Joseph Priestley, to the year 1795'* published in 1806 he plays down his disappointment stating:

> ".. But I was objected to by some clergymen in the board of longitude.....and presently after I heard that Dr. Forster, a person far better qualified for the purpose, had got the appointment. As I had barely acquiesced in the proposal, this was no disappointment to me, and I was much better employed at home, even with respect to my philosophical pursuits." [19]

This second voyage also involved Matthew Boulton. The voyage consisted of two ships *'Resolution'* and *'Adventure'*, Banks was to provide equipment and personnel to assistant him with the scientific element of the voyage. Banks approached Boulton to provide beads, wristlets, earrings, glass pendants, gilt chains and medals which would be given to the natives of lands that Cook might discover.[20] Boulton duly provided the requested material which was put on board *'Adventure'*. Priestley might (at the time) have felt disappointment but so did Banks as the Admiralty's organization was shambolic with Banks' scientific team having little space to work in; and when alterations were made to accommodate Banks the ship was found to be dangerously overweight and thus all the equipment had to be removed! The ships sailed without Banks' scientific team but with Boulton's trinkets.

Like most of the Lunaticks Priestley was personally acquainted with the Reverend John Michell who is constantly connecting people. It is because of Michell that Priestley was contacted by William Herschel. In 1780 Herschel wrote to Priestley asking if he could obtain some information regarding optics and the work being done in this field by Michell. Priestley responded to Herschel:

> "…if you will be so good as to send me an account of the construction and effects of the telescopes you have made, and what farther views you have in the same way, I will not fail to take an early opportunity of writing to him (Michell) to procure an account of what he is doing…."[21]

Michell was contacted by Priestley regarding Herschel's enquiry who eventually replied on 21st January 1781 stating that he was very much obliged to him for the account of what Mr. Herschel had done and what he had seen.[22] Michell duly corresponded with Herschel and recommended Priestley's book on optics.[23] Herschel had already obtained the book to study commenting on certain portions of it, presenting two papers to the Philosophical Society of Bath in 1780.[24]

Priestley was though very much a political animal and away from the Lunar Society meetings he was outspoken. As has already been mentioned Birmingham was more non conformist than many big towns, the established church and the state were always wary of that fact, especially as Priestley wrote books like *'An History of the Corruption of Christianity'*. In words that are recognisable in the political turmoil of the 21st century Priestly commented:

> " We are, as it were, laying gunpowder, grain by grain, under the old building of error and superstition, which a single spark may hereafter inflame, so as to produce

and instantaneous explosion; in consequence of which that edifice, the erection of which has been the work of ages, may be overturned in a moment..."[25]

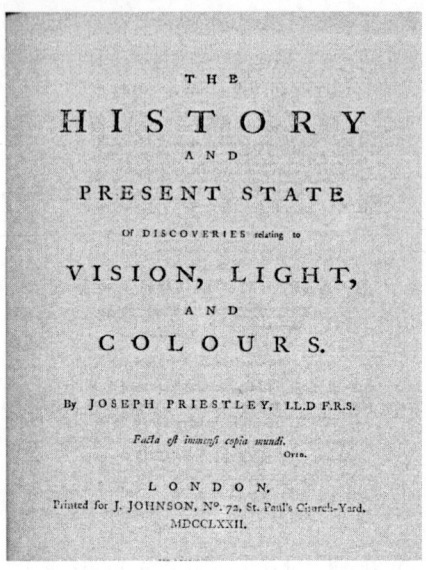

Fig. 31 Priestley's book on optics that was recommended by John Michell to William Herschel. (Lound Collection)

On July 14, 1791 Birmingham became the scene of violent rioting aimed (so it was claimed) at a dinner party to celebrate the anniversary of the storming of the Bastille by French Revolutionaries in 1789, with Joseph Priestley and the Birmingham historian William Hutton as the main targets – even though neither of them were at the dinner.[26] Priestley's views on revolution, and non-conformism were well known, indeed many of the Lunaticks shared the same views to a lesser (and in Boulton's case a much lesser) passion. A mob had gathered outside the *Hotel* in Temple Row but when it emerged Priestley wasn't there they took out their rage on the New Meeting House on Moor Street burning it to the ground. They then attacked the Old Meeting House[27] and the town exploded into violence as meeting houses were destroyed, anyone connected with non-conformism was in danger. William Hutton's Library and home was destroyed. Many joined the mob it seemed, to settle local scores and chaos ensued. Priestley found his house under attack from a 'King and Church' mob, Thomas Wright Hill (who had moved to Birmingham from Kidderminster) offered, with a group of Priestley's pupils assistance to repel the mob. Priestley however refused to meet violence with violence and instead fled with his wife; the rioters broke into his house, some intent on destruction some on looting.[28] Priestley's scientific papers and books were scattered and thrown from windows into the street. Scientific instruments including telescopes lay on the ground next to Newton's *Principia*, a 1614 copy of Kepler's *Paralipomena ad Vitellionem* was trampled underfoot and no doubt a rampaging rioter tore to shreds a 1665 copy of Hooke's *Micrographia*. Priestley wrote to William Withering of his loss:

"...One of the most disagreeable circumstances attending the riot, with respect to my books, is that the sets are almost all broken."[29]

The wanton destruction of Priestley's scientific material by a mob agitated by a combination of church and state officials remains one of the darkest moments in Birmingham's history. Other Lunaticks were in danger too, some had been armed with guns made by Galton and stayed up all night for fear of an attack as the rioting went on for 5 days. Boulton had treated his workforce to a feast to persuade them not to join the mobs. William Withering at Edgbaston Hall found a drunken mob marching on him; he escaped disguised as a wagoner! Two troops of Dragoons arrived from Nottingham and along with light cavalry from Lichfield the riot was quelled. Priestley fled to London producing a book appealing to the rioters of Birmingham, explaining his views. His friends replaced most of his scientific equipment including a telescope in some effort for him to start again and indeed it did not take long before he was conducting scientific experiments. Even so he and his wife emigrated to the United States where he became caught up in politics there, however, on the election in 1801 of William Small's ex-student – Thomas Jefferson, Priestley found at last a government that was friendly towards him. He lived out the rest of his life in the fledgling republic.[30]

Back in Birmingham, a company of soldiers was permanently garrisoned to prevent any repetition of the riots – or perhaps to ensure the population's loyalty[31] - echoes of Prince Rupert. These soldiers would be used again in the next few years although the catalyst was not the French Revolution it would be food prices. The riots had demonstrated that free thinking and speaking individuals were always sailing close the wind. It was time to keep one's head down.

Fig. 32 The destruction of Priestley's House a quarter of a century of work burns in a Birmingham street.(From Samuel Smiles' 'Boulton & Watt')

5. William Herschel's Lunatick Friends

William Herschel is the great astronomical figure of the 18th century with his reputation as a telescope maker, his discoveries of the planet Uranus, two of its moons, two moons of Saturn plus his work on double stars, and the infra-red made him an international figure. Like many people however what they became famous for may not have always been their first career. Music was major part of William's life which was why he came to Britain in the first place. He was joined by his sister Caroline who like her brother was into music performing as a singer with a noted performance in Handel's Messiah at Bath in 1778 which resulted in an invitation to perform at Birmingham's prestigious musical festival - an invitation she turned down.[1] Concentrating working with her brother in astronomy the two of them became one of the great astronomical partnerships in history.

Fig. 33 William Herschel the most famous astronomer in England who formed a close relationship with Boulton and Watt. (From a print in the Lound Collection)

Although Herschel had written to Priestley there is no evidence of further correspondence with other members of the Lunar Society until 1786 when he wrote to Boulton to enquire about a steam engine that Boulton might supply to his friend in Hanover Germany. Boulton and Watt were in France at the time and the letter was passed on to them in Paris.[2] This began a friendship that would last many years with Boulton and James Watt visiting Herschel at Slough on several occasions yet sadly like much of Boulton's correspondence gaps appear. This may be because much of the saved material has some connection with business or engineering. It is however likely that Herschel corresponded with Boulton and Watt irregularly after December 1786 – unfortunately Boulton's and Watt's letters to Herschel do not seem to have been preserved in the Herschel archive either.

In February 1787 Sir Joseph Banks wrote to Boulton mentioning Herschel's discovery or two satellites around Uranus:

> "…Herschell has discovered two satellites revolving around his Georgian Planet. He has not yet completed his 40 feet telescope what new discoveries may not we expect from that?"[3]

Herschel's 40 foot telescope was the talk of society and had drawn special interest from engineers as well as astronomers. King George himself had helped fund the project and Boulton wanted to see it, thus following a trip to London he returned to Birmingham by way of Slough so as to call on Herschel.

Fortunately he recalls the event in a letter to his son Matthew Robinson Boulton (1770-1842) who was in Versailles at the time:

> "...When your sister and I return'd from London we call'd upon Herschall the astronomer near Windsor and saw his great telescopes, the largest of which are fix'd in the open air and weigh several tons. It is govern'd by wheels, by pulleys, and ropes, in such a way as to point it to any of the planets or any of the stars near the eclipffck. The great speculum is full four feet in diameter. The tube of it is five feet diameter and forty feet long, and Mr. Hershall gets into it by a door on side and runs up and down it as easy as upon and down his own stairs. He hath discovered lately two burning mountains or vulcanos in the Moon, and when the largest telescope is quite compleat he expects to make many other important discoveries and extend our ideas of the immensity of creation." [4]

The new correspondence with Herschel as well as visiting Slough may well have sparked something in Boulton as just a year later he began planning a new observatory as part of the redevelopment of Soho House. These developments included adding two wings to the house (Fig 35). The West Wing included rooms for Natural History, Wet Chemistry and Dry Chemistry. The East Wing had a curved room for Botany that led to a domed observatory. The second Soho House Observatory was designed in 1788 by John Rawsthorne (1761-1832) with a dome 16 feet 6 inches in diameter (Fig. 36 & Fig. 37).[5] This was a lavish construction project and would be Boulton's 'Elysium'. It would be interesting to know what type and size of telescope Boulton envisaged placing in the observatory.

Due to a downturn in the economy, costs of the steam engine business and the rising costs of the Soho House additions the plan was abandoned.

Fig. 34 Herschel's 40 foot telescope Boulton and Watt visited Slough on several occasions. (From a print in the Lound Collection)

Fig. 35 Soho House Extensions designed by John Rawsthorne 1788 Boulton's own 'Elysium'. (Birmingham City Archives)

Over the years Boulton became involved in numerous enterprises producing silverware, setting up the Birmingham Assay Office, coin and medal production, and coin press construction. Yet it was the steam engine that took up a great deal of his time, not simply with the design and marketing but with prosecuting those who infringed the steam engine patent. Peter Dollond had been merciless in protecting the patent for the achromatic lens; Boulton and Watt were probably just as active against infringers of Watt's patent. The original patent had been taken out by Watt in 1769 and acting on Boulton's advice it had been a general rather than a detailed patent which like Dollond's patent could leave it open to interpretation which led to various disagreements with other engineers. It was made worse when Boulton and Watt successfully petitioned Parliament for an extension to the patent of 25 years – which was granted. Cornwall was a particular hotbed of dissention with engineers like Jonathan Hornblower Jr.(1753-1815), Jabez Hornblower (d.1814), Edward Bull and Stephen Maberley keen to either copy the engine or begin development from it. Boulton and Watt had considered legal action in 1785 but did not pursue it.

Fig. 36 The Soho House Observatory as designed by Rawstorne.
(Birmingham City Archives)

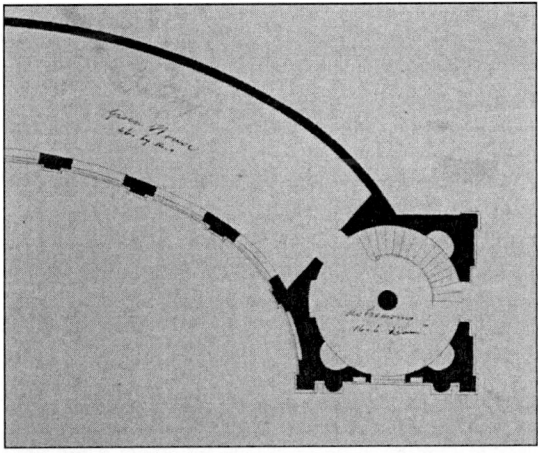

Fig. 37 The Soho House Observatory showing internal arrangement.
(Birmingham City Archives)

Development of converting the reciprocating motion of the steam engine to rotative motion had also taken up time (and money), the solution developed and patented was the 'Sun and Planet Gear' so called as it gives the appearance of a planet rotating around the Sun. This development also called up issues of patent and design priorities. The crank was the preferred option however James Pickard had already used a crank for rotary motion at his mill on Snow Hill in Birmingham and patented the idea in 1781. Boulton and Watt's trusted employee William Murdoch came up with the Sun and Planet gear as an alternative to the crank, a system that Watt patented in 1782. This development increased the demand for steam engines that could now be employed in many roles.

Even with the pressures of business the Lunar Society meetings continued. The meeting of 3rd June 1792 saw the visit of William Herschel with his wife Mary, and godfather General Kimarzewski. They were enroute to Glasgow to receive the Freedom of the City stopping off on the way at various places, taking a sort of *Grand Tour of Britain* acquainting themselves with the places of interest.

Herschel arrived in Birmingham on June 1st where he called in to see his brother-in-law Thomas Baldwin and his stepson Paul Dee Pitt; and then visited Matthew Boulton at Soho House. The next day he met with James Watt, James Keir and Dr. Withering as a prelude to the next evening's Lunar Society meeting where they were joined by Samuel Galton Jr.[6] Herschel had brought with him a 7 foot reflecting telescope – quite a piece of equipment to carry along on such a trip. Was it ever placed on the roof of Soho House? Was it set up in the garden for the Lunaticks to observe the heavens, guided by the country's leading astronomer? There is no written record of what was discussed on that *Lunar Night*, but one can imagine the excited exchanges.

The next day (June 4th) Herschel was taken on a tour of the Soho Manufactory no doubt being suitably impressed. On the 5th Boulton showed him the Soho Mint where he observed the striking of medals. He seems to have had a particular interest in this as he made some detailed notes and calculations in his journal, writing:

> "Saw the coinage Mill strike 54 medals in a minute by one block and there are 4 at work at once and 4 are at rest that at night have been put in motion so that 54 x 8 = 432 large medals per minute or 25,290 per hour or 622,080 per day might be struck off."[7]

Herschel spent another evening with Boulton at Soho and one wonders what subjects they discussed. The timing of Herschel's visit was at a difficult time for Birmingham, as only a year before the King and Church Riots (or Priestley Riots depending on your emphasis) was raging on the streets. Herschel's comments on those are not recorded. From Soho Herschel moved on to James Keir's house at Hill Top. He stayed at Keir's until the 9th, spending the days visiting an ironworks at Bradley detailing his observations on the production of iron and the making of a 50 foot long boiler.[8] Keir escorted Herschel around his own works at Tipton demonstrating the manufacture of soap, alkali, white and red lead. From Keir's Herschel moved on to Dudley where he visited the castle and the lime quarries stopping in the evening at Mr. Finch's house where the 7 foot telescope was put to good use showing Mrs. Finch Jupiter and its satellites.[9] Herschel's tour of the industrial midlands continued as he visited the cradle of the industrial revolution – Ironbridge, and then on to Coalbrookdale where he took delight in a steam engine with a 67 inch diameter cylinder, and then studying in some detail the workings of a blast furnace at John Wilkinson's (1728-1808) works.[10] From there he went on to Bridgenorth, Shrewsbury, Ellesmere, and Wrexham arriving on June 14th at Bersham and another of John Wilkinson's works; here Herschel observed the boring of a 34 inch cylinder for a Boulton and Watt steam engine.[11] Herschel I am sure now realised why Wilkinson was known as 'Iron Mad'. Herschel and his party eventually arrived in Glasgow on June 29th but still continued to visit manufacturers stopping at a glass works.[12]

Herschel met and dined with the intelligentsia of Glasgow and Edinburgh – including Watt's good friends Dr. Joseph Black, and John Robison.[13] He was familiar to both, Black had visited Herschel at Slough in 1788.[14]
On his way back home Herschel called at more places on route including Thornhill where he examined the telescope of the then quite ill John Michell.[15] On July 13th Herschel called in on Erasmus Darwin in Derby where he was shown an unusual geological specimen a block of lead ore in a shell of iron ore.[16] Three days later he was back at Soho House with Matthew Boulton, but on this occasion he was just passing through.[17] Herschel's notes on the manufactories and works he visited present a fascinating and unique glimpse of industry in 18th century Britain.

Boulton and Watt were regular visitors to the Herschel's in Slough, the correspondence that remains shows that a close friendship had developed:

> "The inclosed list was given me by my brother-in-law Mr. Baldwin, and as I had mislaid it when I had the pleasure of your visit at Slough I by leave to inclose it here. I forgot whether you took a memorandum of the copying machine that he wished to have. Please to present my Respects to Miss Boulton and your son, as also to Mr. Watt who I hope is well; Mrs. Herschel joins in compliments with me, and I remain Dear Sir Your faithful & most obedient Servant
> Wm Herschel."[18]

The copy machine referred to is the one developed by James Watt with copy paper specially prepared by James Keir.[19] Watt had quite a good sales record for such equipment and it is thanks to such machines that so many letters of Boulton and Watt have been preserved. The list referred to is missing but it was more than likely a list of items that Baldwin wished to purchase from Boulton. Boulton's response is missing but in July 1793 Herschel writes:

> "By the contents of your letter I find that the Amusements of the Town have no charms for you. It will make your Elysium Soho so much the sweeter. We shall be happy to see you in your way to that blissful mansion. We shall not be at liberty to leave Slough till the latter end of this month and after a little tour to a friend of ours near Warwick we mean to come to pay our respects to you, and hope you will by that time have quite forgotten all the pretty expensive amusements that entertain you now.
> With many compliments from Mrs. Herschel to you and Miss Boulton,
>
> PS. I beg to be remembered to Mr. Watt." [20]

Herschel visited Soho House again in late July 1793 spending three nights before spending sometime with James Keir. The expensive amusements (as underlined by Herschel) may refer to the riots or the dissention among other Birmingham manufacturers due to Boulton offering far better pay and conditions at Soho. Perhaps it was these close ties with Herschel that promoted another burst of optical interest; James Watt purchased a telescope with stand from Charles Lincoln of 32 Leadenhall Street. The telescope came with glasses that could magnify 25 and 50 times.[21] Matthew Boulton had taken an interest in this instrument and made some calculations on the magnifying power of Watt's new 'scope compared with one of his own.[22] It may have been Boulton's comments that led Watt to send the telescope back to Lincoln where modifications were made to Watt's specifications and returned.[23]

Boulton had a habit of comparing telescopes; in August 1780 he was in Cornwall on Sunday 27th along with Mr. Derwin (probably Erasmus Darwin) and Mr. Henderson.

"…we took the telescope and by its assistance we counted distinctly 50 ships off Falmouth."

Two days later he spent the morning with Dick Phillips and a Mr. Inniss comparing Inniss' telescope with his own and found it was better.[24] Along with the Lincoln telescope Watt bought a pocket achromatic which he probably gave to his son James Watt Junior.[25]

Boulton visited Herschel in Slough in March 1794 as can be deemed from another communication from Herschel.

"About two hours after you left Slough a Chaise driver came to me with a sorrowful countenance saying that he was not paid for his job in bringing two Gentlemen from Hounslow, and wished to be informed if I knew them. After a little examination of his face I thought it was an honest one, and therefore I took the liberty of brightening up his looks by paying him the usual fare, and fee for himself.

I hope you got safely to Soho and found all your valuable Family well, and with my best respects to Mr. Watt & all friends." [26]

Boulton it seems failed to pay the cab! There is no record if Boulton reimbursed Herschel, but no doubt he did.

6. Astronomical Witnesses

The relationship with William Herschel was not just social; Herschel along with James Lind, Jean André de Luc, and Jesse Ramsden acted on more than one occasion as expert witnesses for Boulton and Watt in legal proceedings against those who had allegedly breached the steam engine patent. It must be said to walk into any court with Herschel, Lind and De Luc (all of whom were close to the King) as witnesses is going to give Boulton a distinct advantage.

Although Boulton and Watt had previously refrained from full court action, in 1792 they didn't hold back; which ponders the question whether the issue was discussed with Herschel during his first visit to Soho. It was not uncommon for Boulton's friends to supply intelligence information from around the country. John Warltire while lecturing near Radstock in 1783, had spied on Jonathan Hornblower obtaining a drawing of Hornblower's compound steam engine which he duly supplied to James Watt.[1] Jesse Ramsden was keeping an eye on Edward Nairne who planned to make a model of a Watt type steam engine; so it seems that Boulton and Watt operated quite an intelligence gathering network. No doubt Lind and De Luc kept an ear to the ground at Windsor, and certainly Watt's friends in Scotland were looking after his interests there. Watt of course seems more and more concerned about infringements of his patent and I wonder if Maskelyne's attempt to claim Watt's design of a prismatic micrometer years before had heightened Watt's sensitivity to such matters.

Once begun the court actions dragged on, James Watt Jr. acting in much the same way as Peter Dollond – whether Watt saw the irony I do not know. Witnesses were called upon and James Lind accepted on behalf of himself and William Herschel.[2] An astronomer may seem a strange choice for a technical witness, but one must understand that within technical matters of the 18th century an astronomer was seen as a highly respected position, what's more from the point of view of trying to impress a court and jury the country's leading astronomer is going to carry a lot of weight. Yet Herschel's name alone would not be enough, he would need to actually know something about the subject in hand. Herschel was well versed with the steam engine.

His tour in 1792 was not just to take in the sights, it was used as an educational trip and he visited many manufactories and works noting the technical details of machinery. From his diary one can see that he noted the production of boilers and cylinders for steam engines. He had visited John Wilkinson's ironworks at Brosley and Bersham where cylinder's for Watt's engines were made.[3] He also saw steam engines at Coalbrookdale,[4] Manchester,[5] Edinburgh,[6] Sunderland[7] and Truro;[8] he observed ironworks in action complete with equipment such as puddling furnaces at Ketley,[9] and bellows blowing while at Bersham. He had seen Watt's design in actual use and a Newcomen engine just outside Sunderland, so he was able to make comparisons of the use and operation. Herschel had obviously conversed with Boulton and Watt but also with Joseph Black and John Robison in Scotland and I suspect that Herschel's tour was used as an opportunity for him to become acquainted with the engine for his future role as a witness. Certainly Robison and Black would have been able to give him exact information on Watt's developments and may have even shown him the model Newcomen engine that Watt had repaired years before which had started Watt on his separate condenser design. I would also not be surprised if any useful intelligence he had gained on his visit to the various works was not passed on to Boulton and Watt when he returned to Soho on July 18th and August 6th 1792.[11] It is interesting to note from his diary that after returning from Glasgow he toured the West Country with a stop off in Redruth and Truro, one of the seats of opposition to Boulton and Watt, where he visited steam mills.[12]

The initial court case was heard in June 1793 with a jury finding that the patent had indeed been infringed; however, there was a question as to whether the patent was good in law – a point both Small and Watt had argued about Dollond's patent for the achromatic lens. The case regarding Watt's patent was heard in February 1795 with the judges being divided on the issue.

Boulton and Watt had taken other cases to court and were successful in getting injunctions against Hornblower, Bull and others.[13] The validity of the patent was still to be settled in law and this final court drama was to take place in 1796. James Watt Junior acting on behalf of his father wrote to William Herschel:

"Dear Sir,

It is the unfortunate lot of some people to be always troublesome to their friends, and no sooner are they extricated out of one scrape then they get into another. This is precisely the case with your good friends Boulton and Watt for scarcely have they finished their visit in Chanary against Hornblower & Maberley before they are forced into an action at Common Law with these same parties and obliged again to solicit the evidence and the attendance of their friends. A letter which I have this day written to Mr. De Luc and requested him to communicate to you will explain this matter more at length as also the use of certain papers which you will receive per coach. I hope you will be able to attend the trial at Guildhall early next month where the business will come before the Chief Justice of the Commission of Common Pleas & a special Jury of London Merchants. The questions of Law being all disposed of, we shall only have to try 'whether the specification is sufficient to enable a mechanic acquainted with Newcomen Engines to construct one.' The question I hope will soon be disposed of both by the Jury & court & laid asleep for ever.
As there is scarcely time to wait your answer to this letter before the subpoena must be secured of me I presume so far upon your goodness as to suppose you will not object to it. The person who desires it will tender your expenses namely for the sake of extracting us to recover from the defendants.

With best Respects to Mrs. Herschel
I remain, Sir, Your Obliged Servant
James Watt Junior

PS I shall take the liberty of providing lodgings for both you, Dr. Lind & Dr. de Luc in local neighbourhood a day before the trial of which you shall have timely notice."[14]

The case came before Lord Chief Justice Eyre on 16th December 1796. Herschel's friend Jean André De Luc (or John Andrew De Luc as the proceedings refer to him) was called first, he had been a keen student of the steam engine for over forty-five years and in response to a question from Lord Chief Justice Eyre as to whether or not De Luc was acquainted with the use of steam in engines before Watt, Newcomen and Savery's inventions he went into a long description of the nature of steam and how it is employed. On answering subsequent questions De Luc had all the technical figures on the tip of his tongue, talking about the savings on fuel with the Watt design. It seems in essence he was trying to set the scientific baseline to prove that Watt's design is a valid patent. He was examined on Watt's specification article by article.[15] Following the evidence given by De Luc, and a lot of discussion between Lord Chief Justice Eyre and the lawyers, Herschel was called to give his evidence and was examined by Mr. Austruther. The exchange of questions and answers gives an interesting insight into Herschel:

"Mr. Austruther: Are you acquainted with mechanics in general?

Herschel: I am.

Mr. Austruther: Has it been much the subject of your study?

Herschel: Yes, for a great many years, I suppose not less than 30 years.

Mr. Austruther: Are you acquainted with the models in which the properties of steam were used in Engines prior to this invention of Mr. Watt's?

Herschel: Yes for instance how they were used in a Newcomen Engine.

Mr. Austruther: Is there any particular form in which it was of necessity that Newcomen's Engine would be erected?

Herschel: The form will be to justify the occasion and the place where to be erected.

Mr. Austruther: You have read Mr. Watt's specification?

Herschel: I have.

Mr. Austruther: Then taking it that you were acquainted with Newcomen's Engine should you have felt any difficulty in forming an Engine from the specification of Mr. Watt which should have produced the effect of saving fuel and steam?

Herschel: I should not have found the least difficulty the specification is so plain to me that I should immediately have executed the engine.[16]

Herschel was then questioned at length on whether or not given Watt's specification he could build each element of the engine. Herschel on all occasions said he could and sometimes gave a detailed answer. After more discussion Herschel was cross examined by Serjant-Shephard asking about the details of the steam engine and if he felt that Watt's patent was valid.

"Mr. Serjant-Shepherd : I understand from you that that specification explains Mr. Watt's principle.

Lord Chief Justice Eyre: Which principle is – that there is to be a separate condenser –

Herschel: That is one of the things mentioned.

Mr. Serjeant-Shepard: Does not the proportion that the condenser should bear to the size of the cylinder make some difference in the working of the engine?

Herschel: An engineer will soon find out what proportion the parts are to have to each other.

Mr. Serjeant-Shepard: Could he find that out without making repeated experiments?

Herschel: Not without some experiments but I do not think that any machine can be made without some experiments – there is a noble invention – the making of Achromatic Glasses by Mr. Dollond you cannot make them without experiments to ascertain the power of the Glass and yet the invention is excellent and well described.

Mr. Serjeant-Shepard: Then a man must make experiments.

Herschel: That is no more than saying that a man must understand his Business before he makes an engine.

Mr. Serjeant-Shepard: An engineer who understood Newcomen's Engine ever so well would not know the proportion that a condenser which was not in Newcomen's Engine should bear to the cylinder without making some experiments.

Herschel: Suppose he was to make it larger there could be no detriment to it.

Mr. Serjeant-Shepard: Might not an experienced Engineer without any great reprobation of his experience make a condenser at first which would not answer in proportion?

Herschel: I think if he was an Experienced Engineer he would not make one that would not answer but the next Engine he makes he would make a better that is the case in all machines in general."[17]

After a long session of questions in which Chief Justice Eyre (who was clearly not ignorant of scientific matters) intervened to clarify what Herschel was saying Herschel was stood down. Reading carefully Herschel's evidence as a whole I have to wonder whether it might have been an idea for him to become a builder of steam engines! Jesse Ramsden then came to the stand and after being sworn in was questioned by Mr. Rouse:

"Mr. Rouse: What is your profession?

Ramsden: I am an optician.

Mr. Rouse: Were you acquainted with the steam engine as used under Newcomen's invention?

Ramsden: It is a thing that every mechanic understands who understands the Fire Engine at all.

Mr. Rouse: Do you know the invention of Mr. Watt?

Ramsden: I do.

Mr. Rouse: I will only examine to the specification you have read this specification?

Ramsden: My memory is not so good as it used to be.

Mr. Rouse: Be so good as read this specification."[18]

Ramsden suggesting his memory is not as good as it used to be is amusing, he knows the specification of the engine very well indeed, briefings between the group of friends would have taken place and Ramsden himself attended the running of engines as a technical witness such as on 11th September 1778 when along with Mr. & Mrs. Whitehurst, Ramsden witnessed the cycle of a new engine erected at Richmond.[19] Ramsden was temporarily excused while he sat and read the specification. There is another great irony in that Herschel referred to Dollond and the achromatic lens, and Ramsden described as an optician gave evidence in defence of a patent that many considered false as its subject was for what many felt was existing knowledge, given that many years earlier Watt had discussed Dollond's patent with Small as false for the same reason.

The Jury found in favour for Boulton and Watt. Hornblower and Maberley appealed and it was not until 1799 (one year before the patent was due to expire) that the whole issue was settled in favour of Boulton and Watt.

Engineering issues and astronomy are often linked as one is reminded that in the 19th century the then Astronomer Royal George Biddell Airy (1801-1892) became embroiled in the railway gauge issue which resulted in a rejection of Isembard Kingdom Brunel's (1806-1859) 7ft wide gauge for the 4ft 8½ inch gauge we still use today. It also underlines the level of respect that senior astronomers held within society – especially that of the Astronomer Royal.

7. Matthew Boulton and the Instrument Makers

Matthew Boulton's connections with the scientific community was extensive throughout his life as much for personal interest as well as business although business was Boulton's nature as can be gleamed from the archives.

Boulton had customers across Europe and thanks to his partner John Fothergill his European contacts were widespread. On occasion Boulton's customers would request a quality telescope or other mathematical instruments through him which, he then ordered from Dollond, and Ramsden and it seems that Boulton & Fothergill acted as an outlet for both Dollond and Ramsden in Europe such as in 1772 when Dollond was requested to supply two achromatic telescopes, one 3 to 4 feet long the other six inches long both "of the very best kind" with the letter stating:

> "Our Mr. Boulton begs his respectful compliments and desires the above may be such as will do both you & ourselves honour with our friendship as well as recommend a further demand."[1]

Boulton's customers were often wealthy, many influential and all expected the very best service which is why they went to Boulton. Yet this still did not mean that business deals were smooth. In 1776 Mr. Hillaire of Metz ordered through Boulton some instruments, optical glass, tubes etc which Boulton then ordered from Dollond, and Ramsden. In September Hillaire wrote to Boulton complaining that Dolland had not responded to the order or answered his letter to him, in turn Boulton wrote to Dollond to push him and the letter clearly states both Hillaire's and Boulton's dissatisfaction with Dollond.[2] However Ramsden comes under fire as well, as he hasn't delivered the instruments he was supposed to supply![3] In October Boulton wrote again to both Dollond and Ramsden wanting to know where Mr. Hillaire's articles are![4] The issue drags on into 1777 and an exasperated Boulton continues to press the two opticians for delivery dates![5] The correspondence on this matter ends there so one assumes Dollond, and Ramsden fulfilled the order. The delays did not it seems prevent further orders being placed.[6]

It was not uncommon for people to write to Boulton to ask his opinion or to offer their inventions to him to manufacture. A good example is that of Peter P. Burdett (1734/5-1793) a cartographer and draughtsmen living in Rastadt, Germany who in 1777 wrote Boulton a most extraordinary letter so filled with interesting information that I have reproduced much of it here:

> "...I have heard our learned and good friend Doctor Darwin say that in the human mind threesome few points weaker than the rest, Hence our unreasonable affections and Antipathy for particular objects. There are many who cannot indure a cat, & I am not singular in having an aversion to a Parson – deprive puss of her ears and tail & dye her of an unusual colour & she will be no longer offensive to the first mentioned people because she is no longer known to them, now I wonder

if something like this was tried upon a priest & the experiment fairly made if the
success would not be equal. The Reverend W. Ludlam of Leicestershire has lately
written a vulgar illiberal & lying letter to Mr. Dollond the Optician in which the
Holy Inquisitor has condemned to eternal Damnation the author of a new
Mathematical Instrument called Optical Compasses and concerning which his
opinion was modestly requested by Mr. Dollond who at that time was in
possession of a model & short description of the same – It would be losing time to
encounter this mathematical Bully as nothing is so invulnerable as nonsense – Yet
I have diverted myself and a few friends in England by sending them some large
Extracts from my remarks upon Ludlam's letter, which perhaps you may one day
meet with if it will not offend you to find that I can treat a Fellow of St. John's as
I would a Fellow of St. Giles's. Now the main purpose of this letter is to give you
Sir some idea of this instrument & without bespeaking your good opinion by
flattery it is as follows: The well known principles of Hadley's Sextant for taking
angles without the necessity of a stand or pillow has in a vast variety of practice
been of the utmost use to me in Astronomical and Geographical Operations both
at sea & land – to extend the utility of this principle I have therefore made an
instrument like the sketch below which sufficiently explains itself to a man of
your knowledge – I shall remark, that no degrees are divided upon the arch, for
after the angle is taken between two objects the Distance of the points of the
Instrument give the chord of half the true angle upon a Radius of a circle equal to
the length of the Instrument which circle or part of a circle is readily made with
the Instrument itself when the points are extended to the utmost = 60 degrees
which are accounted = 120°. Such an Instrument can be well made for one Guinea
& sold for two – the portability elegance & variety of cases to which it can be
applied in measuring Heights, Distances &c without the fatigue of counting
degrees render it a Proper Instrument fit to be had in all well disposed Christian
Familys – but seriously its military properties in the hand of an Engineer in time
siege &c when all other Instruments are either tedious cumbersome or impossible
a real value and importance as I have had the Honor to prove to demonstration
lately before the Emperor of Germany & some of the first personages for Quality
and learning on the continent (to use the language of Card Taylor) – I shall
conclude therefore in saying that if you may Dear Sir have leasure and inclination
to look farther into the Optical compasses you may be satisfied with a view of a
model and short description now in the hands of Mr. Newman No. 3 Sun-Court
Cornhill London. After which should you finde it to have merit enough to be
patronised and worthy to be Ranked with your one wheel clock or multiform S (to
which it has no small resemblance). The Property of the Optical Compasses is at
your service – But upon these conditions that should the profits arising from the
sale of this instrument enable you to add an other manor to your estate that I (too
poor to come under the benefit of the act) the old original inventor shall have full
liberty to shoot & hunt upon the same.
Mrs. Burdett gives me her compliments to your Lady – I beg to be kindly
remembered to Mr. Eggerton and his brother & if any services one in my little
power which can be exerted to your or their advantage in this part of the world
you have only to command

Dear Sir. Your obedient & Very Humble Servant
P.P.Burdett"[7]

Burdett may or may not have known that Boulton had had communication with William Ludlam (1717-1788) when in 1775 Ludlam had supplied to Boulton details of a clock with a wooden pendulum the design of which Boulton was having made by George Donisthorpe (d.1802) of Birmingham.[8] There is nothing in the records that show what (if any) was Boulton's response but he received a further enquiry about the compass from William Law & Neuman who had the model of Burdett's device in their possession.[9]

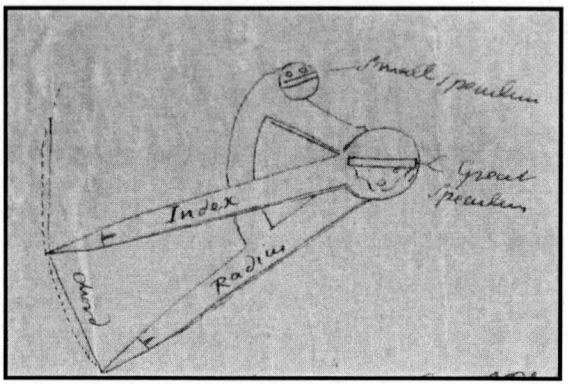

Fig. 38 Burdett's Instrument. (Birmingham City Archives)

Jesse Ramsden is seen by most as the leading instrument maker of the day, and Boulton had a personal friendship with him, Ramsden's wife Sarah (the youngest daughter of John Dollond) acted on behalf of Joseph Poli the Preceptor to the Hereditary Prince of Naples who wanted to purchase a steam engine for the King of Naples.[10] Sarah Ramsden considered Poli a personal friend and from her letters to Boulton it seems she was on very good terms with the Neapolitan Royal family.[11] Her involvement resulted in a large pumping engine being supplied which cost £1000.[12]

Apart from purchasing scientific instruments and components for steam engines from Ramsden, Boulton also supplied brass tubing to him for use in instruments and supplied technical advice. On 29th December 1781 Ramsden wrote to Boulton :

"Having occasion for a very considerable quantity of plated copper I shall esteem it a great favour if you should inform me of the best mode of   procuring them. The money to be paid on delivery. A friend of mine with whom I am concern'd is taking out a patent for drawing tubes for optical purposes plated with silver or gilt. The patent I expect will be out in about 14 days or a little longer and shall have the whole of the tube drawing in this metal by this you will judge if the patent succeeds. Of the quantity that will be wanted and am determined to make it a (illegible) money article in buying as well as selling I must further beg you will be so good as inform me what the plated metal such as you will judge prosper for this purpose." [13]

The friend Ramsden is referring to is Joshua Lover Martin (Benjamin Martin's son) who was granted a patent in 1782 for '*A New-invented Art of Drawing Tubes*

Plated or otherwise covered with silver or gold, on copper or other metal, for the
purpose and construction of telescopes, perspectives, opera glasses, and various
optical, mathematical, and philosophical instruments, to which they are
adapted.'[14]

Martin had contacted Ramsden who in turn contacted Boulton for his opinion.
Boulton was a noted expert on the acquisition of metals – his partnership with
Fothergill had opened up contacts across Europe with Soho manufactory
requiring large quantities of metals. Boulton was also in touch with numerous
mine owners, it was these people to whom he sold steam engines to act as pumps.
He was also an expert on plated goods. About a year later Peter Dollond was
advertising telescopes with brass draw tubes, it has been suggested that Dollond
bought the patent from Martin although there is no firm evidence for this.[15] Draw
tubes par se were produced by various makers[16] however Martin's patent is
specifically for plated draw tubes, and from Ramsden's letter he seems to think
that Boulton may be able to produce them or arrange supply of the metal.

On January 11th, 1782 Ramsden requested some brass tubes to a very exacting
specification and requests that Boulton keep the arrangement secret stating:

> "I have hereinclosed a piece of plated metal such as I have used in my
> experiments and which seems to succeed very well, be so good to send one 3
> tubes of a sort turn'd up and solder'd, I think you can do that just better than I can
> here in Town , I wish this business to be kept secret as possible for some time,
> some difficulties having happened to friend of mine that is concerned with me. I
> believe the demand will in a little time be very considerable indeed, the tubes be
> abt 5 inches long each, and the circumference of the different sizes are 3 to be the
> whole length of the enclosed piece three to be from the middle notch to the end in
> circumference in all 9 tubes pray let me have as soonest possible."[17]

There was much secrecy between instrument makers regarding suppliers and
products a characteristic that is common to this day. Ramsden considered Boulton
not just a supplier of parts and a customer but also a friend. His connection with
Boulton (and Watt) offers a new insight into the working practice of Ramsden.
Boulton and Watt were interested in micrometers and dividing engines and
exchanged correspondence with Alexandre Tournant on the subject, and I wonder
how much information Ramsden gleamed from them on those subjects. Tubes
and cylinders were of great interest to Boulton and Watt, both (and especially
Watt) wanted the most accurate construction especially for steam engines – the
more accurate the cylinder the less chance of leakage or structural failure. Soho
Manufactory could easily produce high quality tubing to Ramsden's requirements
and meet the delivery schedule. Like all industry of the time (and still today)
specialist work may be outsourced to other firms. One Lunatick who offers a
prime example of this is Samuel Galton Jr. He was a keen amateur scientist and
submitted a paper about prismatic colours to the Royal Society in 1782.[18]

Galton was a gun maker in Birmingham working in his father's business (an
interesting occupation for a Quaker!). The gun trade in Birmingham was a highly
skilled industry and guns were shipped to all parts of the world. Guns would often
be sent in kit-packed form to gun shops in various cities, the shop would then add
their own stamp on the gun. So although a gun may be stamped *'London'* it might

actually have been made in Birmingham and shipped to London. Boulton would have done the same for specialist parts for instrument makers; he would produce the tubing and send them out Ramsden and other makers who would then complete their instruments and add their mark. Boulton provided tubes, and plated metal (possibly for Martin's patented draw tubes) to Ramsden as a bill of 5th October 1782 testifies.[19]

In the event tube drawing as suggested by Ramsden does indeed become a profitable enterprise one in which Boulton becomes involved. A machine for the drawing of tubes is ordered by Boulton from – Jesse Ramsden, possibly around 1783 as a letter of June 7th 1784 from John Hodges of Soho Manufactory is chasing Ramsden for the equipment:

> "Just before Mr. Boulton's departure for Cornwall he told me he expected some time ago a machine for drawing of tubes which you had promised to send off immediately – I doubt not that you have delivered it to the carrier but as it is not yet come to hand must beg you will make the necessary enquiry after it and favour me with a line in return..."[20]

Ramsden was noted for his late deliveries on orders the most famous is probably the 23 year delay in supplying a large telescope and circle to the Dunsink Observatory! [21] Ramsden's problems in meeting order deadlines may have been a combination of a lack of skilled workmen to complete work, delays in designing instruments or perhaps Ramsden found that he had bitten off more than he could chew when accepting an order for a large instrument. On September 6th, 1786 Ramsden wrote to Boulton asking for his advice on how to construct a dividing engine:

> "Dear Sir,
>
> Finding myself under some difficulty in the making of a dividing engine I take the liberty to request your advice on that subject only sensible of your good disposition towards forwarding whatsoever may tend to the advancement of the manufacture will forebear troubling you with any apology.
> The prodigious exactness with which we divide mathl instruments (illegible) constructed the engine for that purpose makes me wish to make some of a much larger radius than those at present, but this incurs a very great difficulty in getting that part or edge of the circle in which the teeth that the (illegible) works in are made to be perfectly round, I fear also that the inertia from the weight of a circle of that size consistent in motion may be rather unfavourable from a little experiment just made I am inclin'd to think that very good cast iron would bear cutting with a screw, if so the circle might be cast much lighter with the same strength than could be done with brass it would also expand much less; pray have the goodness to favour me with your opinion and if you think it will succeed persistent to know where you think it would be cast in the best manner I do not regard the expense I would wish to have it 5 or 6 feet diameter. I am also oblig'd to yet solicit another favour if you think it possible to have brass made in England sufficiently long for a quadrant nine feet radius the Bristol Company with whom I (illegible) dealt and who on many occasions have done everything in their power to oblige me tell me that it cannot be done here & that the brass

makers can have no spirit to put themselves out of the way to improve their manufacture, seeing that the combination of the anglesia and Cornishmen may very soon make it impossible for the brass makers to carry on their business, most probably in a short time, they say that the supply of London will be from a warehouse only and they may under a necessity of declining business. In this state I should esteem it a singular favour if you send confirm me where is the best chance of getting it done I would willingly give the price of British (illegible) if it can be had good and as this size might be out of the common size would pay 50 or 100 £ on the beginning to undertake it and the remainder when it is ready to be delivered pray excuse this liberty.

And I am most respectfully
Sir, Your most obedient servant
J. Ramsden ".[22]

Ramsden has clearly hit a problem and with the tone of the letter he is in need of a quick solution. Given the date it is likely the machine he is looking at is one connected with the construction of an altitude and azimuth circle for the observatory at Palermo founded by Giuseppe Piazzi (1746-1826). This instrument had a five foot vertical circle, Ramsden and Piazzi had problems with it abandoning construction twice before finally completing it in August 1787.[23]

The problems he has obtaining the material for the nine foot quadrant demonstrates a common problem with the improvement and development of techniques outstripping manufacturing capability, a situation that would continue into the 19th and 20th centuries.

It was not just with material and advice that Boulton supplied Ramsden. Workmen too were supplied. In the late 18th century anyone having the credentials that they had served their apprenticeship at Soho Manufactory could almost guarantee employment at any top manufacturer. The high quality of craftsmanship that Boulton and Watt had insisted upon for the manufacture of steam engines and at Boulton's manufactory had the effect of improving the quality of workmanship throughout the brass trade. Ramsden was particularly pleased with a man called Kelly who: "...is a man who is very useful to me and should be glad he would return his mother who I find is a very bad Woman and a great torment to him..."[24]

Another man called Webb who works at Soho manufactory is also referred to in positive terms "I would gladly employ him tho' a rough hand he was useful enough that he is always in some mischief or other."[25] Webb duly arrives in London mistakenly thinking he is going to work for Edward Nairne, Ramsden employs him.[26] Boulton would get great satisfaction from this as he was suspicious that Nairne was going to make a model steam engine to the Watt design with a separate condenser breaching Watt's patent. Ramsden was going to keep an eye on Nairne and let Boulton know if he did breach the patent.[27]

William Baddily was yet another workmen who arrived at Ramsden's from Soho, he brought his wife and family to live in London and needed some assistance, Ramsden asked Boulton to give Mrs. Baddily 4 Guineas on his account.[28]

Fig. 39 Jesse Ramsden, behind him is the Palermo instrument and next to him the dividing machine that Boulton assisted him with.
(From a print in the Lound Collection)

Boulton and Ramsden remained in contact up to Ramsden's death in 1800. A diary entry by Boulton in 1798 reads:

"Mr. Hollywell lives at No.1 Rosamund Street, Charkenwell. Did work with Troughton also with Weeks[29] and is a very good workman and a man of talents recommended by Ramsden."[30]

Workmen were recommended to Boulton by Ramsden and other instrument makers demonstrating just how free moving skilled workmen were in 18th century England.

Tubing was not the only astronomically related products to be produced at the Soho Manufactory. Boulton manufactured fabulous casings for clocks and assembled the complete articles at the manufactory. Many of these clocks had design input from another Lunatick – John Whitehurst the famous clockmaker.

Whitehurst had been a long time friend of Boulton who had visited him in Derbyshire on many occasions. Although known as a clock maker, like all of the Lunaticks he had a wide range of interests including meteorology, geology, mechanical engineering, hydrology and astronomy. It is felt by Maxwell Craven[31] that he was the anonymous author of astronomical articles in the *Derby Mercury* reporting in 1781 that a new comet discovered by William Herschel of Bath should be passing in front of the Sun. Apart from Boulton, Whitehurst had a strong friendship with Erasmus Darwin who re-launched the *Derby Philosophical*

Society in 1779 another circle of people interested in science in which Whitehurst was closely associated. Whitehurst also had a strong association with James Ferguson working with him on the design of several clocks.[32] When James Ferguson gave his lectures in Derby in 1762, 1764 and 1771 he demonstrated his orrery which would have greatly interested Whitehurst. Whitehurst himself was a promoter for the 1764 lectures held at the County Hall on 9th July at 6pm.[33] The connections between Whitehurst and Boulton with Ferguson always remained strong, Ferguson and Boulton had remained in contact since 1761 with Ferguson providing advice on clock mechanisms, time formulae and equations for use in the construction of clocks, and working with John Whitehurst, Boulton designed some beautiful examples of ornamental clocks. Whitehurst had made several astronomical clocks and in 1767 Josiah Wedgwood wished to purchase such a clock from him.[34] Of particular interest in this study however are two astronomical clocks one of which remains on display at Boulton's home – Soho House.

The first is known as the *Geographic Clock* and features a globe of the earth mounted on the top which tilts according to the season. A golden Sun is mounted on a rod which shows the position of the Sun over the earth. Ferguson gave advice for this clock and was asked to provide the globe. Ferguson had been a producer of globes however by this date he had stopped this work but he did source the globe for the clock from Nathaniel Hill. [35] The globe was however supplied 7/8th of inch too small which meant that Whitehurst had to remake the horizon ring.[36]

The second clock was first conceived by Boulton in 1771 as a companion to the Geographic Clock. Again it was a Boulton-Ferguson-Whitehurst collaboration. Boulton wrote to Whitehurst:

> "I think I shall make another with a celestial globe on the top and turned by a clock to show the Sun's place and what stars are upon the meridian."[37]

The timing of the suggestion is interesting, as it ties in with Ferguson's second series of lectures in Birmingham. Whitehurst, Boulton and Ferguson found time to get together to discuss the ideas surrounding the Astronomical clock with Ferguson spending sometime at Soho House and Whitehurst travelling to Birmingham from his home in Derby.

Ferguson produced some drawings to assist Boulton's workmen and this information was passed to Whitehurst through Boulton's clerk. Boulton wrote to Whitehurst:

> "We shall want another exceeding good clock something like (the) drawing Mr. Ferguson gave you but with a sidereal face in the middle."[38]

The clock's face features a silvered star plate as described by Boulton in a letter to Whitehurst:

> "I mean to send to you a piece of copper plated with silver for the star plate which when you have hammered flat and mounted upon its proper axis and stoned fine with fine pumice and then with charcoal so as to have a true surface, I will then be the favour of Mr. Ferguson to lay down the stars and circles truly upon it and you

may engrave the names of some of the first rate stars contagious to them and then drill a small hole in the place of each star will serve us to rivet our gilt stars, which stars we shall stamp to three sizes also grave the months and days..."[39]

*Fig. 40 John Whitehurst clockmaker and possibly a writer of astronomical articles.
He supported James Ferguson's lecture programmes in the Midlands.
(From a print in the Lound Collection)*

This star plate revolves within the chapter ring and is engraved with the day of the month, the month, the names of the signs of the zodiac, a 360° scale, circles representing the Tropic of Capricorn, the Equator, the Tropic of Cancer and the Arctic Circle along with the major stars and constellations. The star plate was originally set for London but was redrawn for St. Petersburg, Russia where the clock was to be sent. A great deal of the technical design work of the clock was done by Ferguson. With Ferguson and Whitehurst working on the technical aspects of the clock Boulton worked on the lavish ormolu case. The sidereal and the geographic clock went up for sale at Christie's in 1772, but neither sold to the bitter disappointment of Boulton. The sidereal clock was then adjusted for St. Petersburg and a description of it was sent to Empress Catherine of Russia. Boulton's partner Fothergill thought the clock might fetch £275,[40] he was wrong, there seemed little interest in it, and Boulton suddenly seemed reluctant to part with it. His partner John Fothergill bemoaned the lack of interest in the clock:

"I despair of ever being able to dispose of the sidereal clock at the price it ought to bring us. It is my wish and I do not think it most advantageous for us to present it to the Empress of Russia..."[41]

Fig.41 The Geographic Clock, the globe
mechanism demonstrates the seasons.
(Private Collection)

Fig.42 The Astronomical Clock at
Soho House, a joint effort by Boulton,
Whitehurst and Ferguson..
(Birmingham Museums & Art Gallery).

Eventually in 1776 the clock was sent to the Empress Catherine (1729-1796) and her lover Prince Grigori Potemkin (1739-1791) both of whom thought it was a fine piece of workmanship but as it did not strike the hours or play music, they would not buy it.[42] Mr. Fordyce wondered whether it might be raffled or be the prize in a lottery.[43] The clock was returned to Soho in 1787 on board the ship 'Nottingham,' minus its glass case which had been damaged.[44] Boulton referring to it as the *'Star Clock'* placed it in Soho House. It spent some time at Great Tew, but it is now back at Soho House on public display. The clock remains in my mind the most beautiful sidereal clock in the country.

Another astronomical theme Boulton used for a piece of art was *Urania the Muse of Astronomy*. This theme was used for a beautiful pocket watch stand (Fig. 44). The figure is bronze and holds a timepiece against an obelisk of statutory marble in the pedestal of which is an enamelled tablet showing the equation of time – supplied by James Ferguson. To the corner of the base is a celestial globe.[45]

Fig. 43 The Star Plate of the Astronomical Clock at Soho House designed by James Ferguson. (Birmingham Museums & Art Gallery)

Fig. 44 Urania Watch Stand by Boulton and Fothergill c1777. (Private Collection)

8. Boulton's Astronomical Twilight

With the steam engine patent nearing its expiration and both men having many other interests, their sons Matthew Robinson Boulton, James Watt Jr. and Gregory Watt would take over a reformed company; this included the construction of *Soho Foundry* in order that complete steam engines could be built to the finest exactitude. The Foundry was completed in the winter of 1795/6 and a small section of it - namely some of the workmen's cottages - is still standing.[1]

At this time Boulton brought in James Wyatt to look at modifying his house. Wyatt had prepared great plans including a radically extended frontage; however this was abandoned even though some work to accommodate this had already been carried out. This work included a new roof structure which was re-faced in slate by Samuel Wyatt and given a new south face elevation. The redesigned roof is the one that is present today, which is a quadrangle, with four hipped gabled runs joining at the corners to leave a small roof flat in the centre which was used as a telescope platform. The platform on the roof has a dedicated flight of stairs and there is ample space to mount telescopes. The viewing aspect is directly to the south as the house has been aligned to the compass points.[2]

Fig. 45 Staircase up to telescope platform on roof.
(The Planetary Society (UK), Soho House Roof Survey 1999)

There is a contemporary (1798-1801) description of Soho House that mentions the roof. Stebbing Shaw surveyed Staffordshire and wrote of its history and antiquities. In his chapter on Handsworth he writes of the Soho Manufactory, Soho Foundry and of course Soho House where he comments:

> "At the top of the roof, which is made very neat and commodious, either for common or telescopic observations..." [3]

Boulton essentially retired in 1800 and began astronomical observing again purely for pleasure. He purchased a new equatorial telescope from Nairne in London.[4] In retirement he was returning to his interest of thirty years previous. Astronomy had moved on somewhat since his interest began. Herschel's discoveries had stimulated a wide interest in the subject and the possibility that other planets remained undiscovered led to many searching the heavens. On January 1st 1801 Giuseppe Piazzi discovered what at the time was referred to as a new planet orbiting between Mars and Jupiter. It was named *Ceres*, and later was redesignated an asteroid and has since been reclassified as a dwarf planet. It was for Piazzi that Boulton many years earlier had aided Ramsden in constructing two large instruments.

It is apparent that he was an ardent observer as in 1801 in a letter to Olaus Warburg a professor of astronomy at the University of Copenhagen he comments that he is unhappy that the political situation between Britain and Denmark is strained; he would however like to communicate on scientific matters:

> "...I will not touch upon political subjects, but if anything new in science should arrive I shall be happy to preserve a mutual communication. I have lately purchased a compleat Equatorial Instrument but having caught an astronomical cold I mean to reserve those Nocturnal amusements till a warmer season." [5]

Fig. 46 Telescope platform area, note skylight in the centre. (The Planetary Society (UK), Soho House Roof Survey 1999)

His observations would have been of the alignment of Mars, Jupiter, Saturn and Uranus which although interested him did not please his daughter Anne who was staying with the Dumergue family in Piccadilly where she had spent Christmas. She was concerned that her 73 year old father was spending too much time on the roof of his house in mid-winter:

"My Dear Father,

My Brother and your letter arrived nearly at the same time and I am much concerned to learn from both that you are unwell. The season is too severe for astronomical observations and we all petition you to lay aside the telescope 'till warmer weather." [6]

He certainly was going to keep observing, he bought yet another telescope in October 1801 of Webb paying £4 14/6.[7]

Fig. 47 Soho House with telescope platform three telescopes are shown marked A,B and C (Artwork by author)

William Herschel paid more visits to Soho on August 3rd 1801 and September 2nd 1806 [8] and one wonders if he stood on the roof of the house and made some observations with Boulton and Watt.[9] Herschel was at Watt's on August 4th 1801 before visiting James Keir on the 5th.

During the 1806 visit William and John Herschel visited Soho Foundry and saw the "inflammable air illumination" this was William Murdoch's gas lighting. John wrote in some detail a description of a gas chandelier that hung in Soho Foundry.[10]

Watt retired to his house at Heathfield not far from Boulton and concentrated on his pet projects such as making a machine to copy busts. The old firm became known as Boulton, Watt and Company and it was the sons who ran it.

The Lunaticks continued to meet even as their number reduced. John Whitehurst died in February 1788, Josiah Wedgwood 1795 followed by William Withering in 1799.

Erasmus Darwin who had done so much to correctly foresee the future of optical systems died in 1802. It is Darwin whose incredible mind helped fuel the fires of enquiry that is the hallmark of the Lunar Society. Desmond King-Hele has written several books which detail the work of a man who mixed science and art with his poetry to celebrate the ideas of Herschel and John Michell. In 1783 Michell suggested that some stars could become so massive that light could not escape them – a Black Hole in modern parlance.[11] Darwin's prose is as dramatic now as it was then:

> Star after star from Heaven's high arch shall rush,
> Suns sink on suns, and systems crush,
> Headlong, extinct, to one dark centre fall,
> And Death and Night and Chaos mingle all![12]

Darwin even expanded on Michell's ideas of black holes suggesting a Big Bang theory for the formation of the Universe when he stated:

> "It may be objected that if the stars had been projected from a Chaos by explosions, that they must have returned again into it from the known laws of gravitation; this however would not happen, if the whole of Chaos, like grains of gunpowder, was exploded at the same time, and dispersed through infinite space at once, or in quick succession, in every possible direction. "[13]

He is also suggesting that the universe may well expand forever. The *Hubble Space Telescope* has provided some evidence that this is indeed the case, with the expansion of the universe actually speeding up! Darwin was not just a loss to the Lunaticks, he was a loss to the world.

Matthew Boulton died on 23rd August 1809 and was laid to rest in Handsworth Church. The Lunar Society essentially faded away. James Watt lived for another ten years with astronomy entering his life once more. He remained in contact with the Herschel's with William and John visiting Watt on July 17th and 18th 1810.[14] In 1813 The *Glasgow Society for the Promotion of Astronomical Science* was formed and they wished to build an astronomical observatory for public use. Professor James Mylne (1757–1839) wrote to Watt to ask for support. Watt although not wanting any official position in the Society was willing to make a donation of £50 – quite a sum in 1813. Mylne replied to Watt including a sample of the minutes from the Society's meeting:

> "An extract of a letter was read from James Watt esq. of Birmingham addressed to Professor Mylne being an answer to an application from the professor to that gentleman to know, as a member of their society in which Mr. Watt states that although he declined entering into any association he was desirous of encouraging

this observatory as a public benefit and was willing to make it a donation of £50. So soon as a sufficient number was subscribed to carry on the undertaking in this meeting duly sensible of the patriotic generosity of Mr. Watt in making this offer. (Illegible) thereof on these terms and most unanimously return him this nominal thanks for this very disinterested donation. They do also resolve and agree that Mr. Watt shall be admitted as honorary member of the Society with the same benefit to his heirs & that he and they shall be entitled to the privilege of visiting and frequenting the observatory in all time coming."[15]

Such generosity to encourage the promotion of astronomy was in fact typical of Watt; he made several anonymous donations to various organizations.

In 1819 one of Watt's last visits was to William Herschel as noted by Caroline Herschel:

"(June) Mr. and Mrs. Watt dined one day with us at Mr. Beckwith's. It was the last time the two old Friends did meet, for Mr. Watt died soon after."[16]

Watt died on 19th August 1819 and was laid to rest close to his business partner Matthew Boulton.

William Herschel's son John remained in contact with the Boulton and Watt families well into the 1820s writing to request a visit which, although general visiting to the Manufactory and Foundry had stopped due to industrial espionage, Matthew Robinson Boulton granted one for Herschel.[17]

Fig. 48 Steam Engine with Sun & Planet Gear transferring Reciprocal to Rotative Motion for which Boulton, Watt and Murdoch are famed throughout the world. (From Smiles' 'Boulton & Watt')

9. Concluding Remarks

It was just a minor enquiry into a statement made by James Keir in 1809 that led me to investigate the Lunar Society and their astronomical interests; it has revealed a fascinating number of links between science enthusiasts, instrument makers and astronomers that is only partly understood. I will endeavour to continue the research. What has been discovered is that Matthew Boulton was a man who had enormous energy and an enthusiasm for knowledge. Having been a successful manufacturer, gained wealth through two marriages and from his late father, by 1770 he could afford to spend a great deal of time and money on his hobbies with astronomy becoming a major interest with plans for a fully equipped observatory. Indeed his friends also seem to have had a common interest in astronomy at this period with active experimentation into the improvement of optics and the extensive correspondence between Lunar Society members sheds light on the path Matthew Boulton was taking. A natural net-worker he made connections across Europe exchanging ideas and assisting where he could. He was, with William Small and Erasmus Darwin the founder of that remarkable

Fig. 49 Matthew Boulton in Later life. (Birmingham Museums & Art Gallery)

group of friends who referred to themselves as the 'Lunaticks' a group whose own links with the engineering and scientific community led to a phenomenal exchange of ideas many of which were used practically in the various industries and businesses of the members. All appear to have a link in one way or another to John Michell, who not only encourages but also introduces people of similar interests together. Astronomy wise, Watt, Small, Boulton, Whitehurst, Wedgwood, Darwin and Keir are all involved in the development of instrumentation passing on their knowledge to others who benefit greatly. Ramsden benefited greatly from Matthew Boulton's engineering skills coupled with his knowledge, expertise, manufacturing facilities as well as highly skilled workmen from Soho Manufactory and his good connections. I wonder just how involved members of the Society became in his projects. Boulton thrives on his enthusiasms and his mind can be stimulated by intellectuals and craftsmen. James Ferguson is certainly a great influence, in 1761 and 1771 his lectures appear at the start point of Boulton's astronomical endeavours; however one cannot view the excitement of science in 18th century Birmingham without including the influence of the itinerant lecturers who provided a valuable service in bringing the subjects of Astronomy, Chemistry, Mechanics and nature to the general public. The tradition of such activity continues to this day in Birmingham with the likes of the *Odyssey Class Dramatic Lecture.*

Fig. 50 Soho House as it is today open to the public as a museum.
(The Planetary Society (UK), Soho House Roof Survey 1999)

In the real world of industry there is rarely a straight line of development, of one person or firm making and developing everything in-house. It is always through connections, out-sourcing expertise, borrowing or even stealing knowledge and techniques that lead to great developments. In the Lunar Society we see a group of men who were part of that whole development structure of the 18th century that would push forward the development of science and industry that would lead to our modern world. Their effects on instrument makers was more profound that I (or anyone save for Eric Robinson) could imagine. Further research in this field would be beneficial to help us understand the networking of scientific ideas and development of instrumentation including the free movement of skilled workmen from one firm to another. It would be interesting to see how the transfer of scientific knowledge across the country was assisted by the various Lodges of the Free Masons as the Lunar Society members in most cases were members of various lodges.

History is full of what ifs. The study into the Lunar Society and their astronomical activities has thrown up a remarkable what if. Encouraged by John Michell and James Ferguson, Boulton was on the brink of becoming an amateur astronomer in much the same way Herschel developed. However with the financial problems of Roebuck and the constant encouragement from Dr. Small, Boulton became involved in the development of the steam engine and thus his money and time became more directed to the new partnership with James Watt. Astronomy for Boulton and indeed the other Lunar Society members was placed on a back burner. Watt himself was deeply involved in instrument making and working with the fledgling Glasgow Observatory seemed to be more in the line of Ramsden than Newcomen, yet as an instrument maker and repairer he was given that model of a Newcomen engine and he moved in a new direction; one that would bring him into partnership with Boulton. Even so Boulton always kept in touch with astronomy and optics and even planned a new large observatory in the 1780s, but again circumstances prevented him from pursuing this endeavour. His time and money were spent fighting patent infringements, developing new markets for his manufactory and developing his coinage enterprise.

With his and the Lunar Society's experiments in optics Boulton was on the brink of moving from the past-time astronomer to a gifted amateur. What if Roebuck had not had financial problems or if Watt had been supported by another? Matthew Boulton may never have become so involved in the steam engine and today he may well have been known not as the partner of James Watt but as an 18th century astronomer and Birmingham may well have had two marvellous 18th century astronomical observatories.

Fig. 51 How Boulton may have been remembered had it not been for Roebuck's bankruptcy. Standing by a telescope holding a copy of Ferguson's 'Astronomy Explained...' The telescope shown here is similar to the one that he had in his observatory, a large aperture Dollond refractor with an equatorial mounting.
(Artwork by the author)

APPENDIX A
Philosophical (Science) Lectures in Birmingham in the 18th Century

Start Date	Description	Lecturer	Venue
2 Aug 1742	Display of Opticks	Unknown	*Wheat Sheaf* Bull Ring
22 Nov 1742	The Microcosm	H. Bridges	*Old Cross* High Street
26 Mar 1744	Electricity	F. Midon	*Beckett's* New Street
26 Apr 1744	Electricity	F. Midon	*Beckett's* New Street
10 May 1744	Air & Electricity	F. Midon	*Beckett's* New Street
28 Jun 1744	Air & Electricity	F. Midon	*Beckett's* New Street
17 Mar 1746	Optical Machine	T.G.	*Spread Eagle* Spiceall Street
Oct 1746	Electricity	Mr. Smith	*Old Cross* High Street
Dec 1746	Electricity	T. Yeoman	*Saracen's Head* Bull Street
Jan 1747	Electricity	T. Yeoman	*Saracen's Head* Bull Street
10 Aug 1747	Philosophical	B. Martin	*Taylor's* Cherry Orchard
17 Aug 1747	Philosophical	B. Martin	*Taylor's* Cherry Orchard
15 Aug 1748	Lectures	S. Hayes	*Woolpack* Moor Street
8 May 1749	Philosophical	J. Arden	*Packwood's* Cherry Orchard
July 1749	The Microcosm	H. Bridges	*Packwood's* Cherry Orchard
Aug 1749	The Microcosm	H. Bridges	*Packwood's* Cherry Orchard
11 Dec 1749	Philosophical	J. Arden	*Packwood's* Cherry Orchard
23 Apr 1750	Philosophical	B. Martin	*Packwood's* Cherry Orchard
30 Apr 1750	Chemistry Geology	Unknown	*Packwood's* Cherry Orchard
June 1754	Microscope	Unknown	*Price's* High Street
April 1755	Chemistry Mechanics	W. Griffiss	*Packwood's* Cherry Orchard
May 1755	Chemistry Mechanics	W. Griffiss	*The George* Digbeth
July 1756	Wool Machine	Mr. Nevil	*White Hart* Digbeth
Aug 1756	Wool Machine	Mr. Nevil	*White Hart* Digbeth
Sep 1756	Philosophical	W. Griffiss	*Old Cross* High Street
May 1757	Astronomy course	D. Silk	Digbeth
June 1757	Philosophical	J. Hornblower	*Jesson's Court* High Street
Dec 1760	Science Night School	T. Hanson	*Hinckley's School*
Aug 1761	Philosophical	J. Ferguson	*Assembly Room* The Square
24 Aug 1761	Philosophical	J. Ferguson	*Assembly Room* The Square
7 Sep 1761	Philosophical	J. Ferguson	*Assembly Room* The Square
Sep 1761	Philosophical	J. Ferguson	*Assembly Room* The Square
5 Jul 1762	Mechanical Display	R. Hawkins	*Hawkin's* 65 Edmund Street
Dec 1762	The Microcosm	H. Bridges	*Red Lion* Bull Ring
Jan 1763	The Microcosm	H. Bridges	*Red Lion* Digbeth
10 May 1765	Philosophical	J. Arden	*Assembly Room* The Square
July 1765	Philosophical	J. Arden	*Assembly Room* The Square
Sep 1765	Philosophical	J. Arden	*Assembly Room* The Square
Apr 1767	Philosophical	J. Arden	*Assembly Room* The Square
Oct 1768	Anatomical	Unknown	*Cooke's* Cherry Orchard
22 May 1771	Philosophical	J. Ferguson	*Assembly Room* The Square
10 Jun 1771	Philosophical	J. Ferguson	*Assembly Room* The Square
1 Jul 1771	Philosophical	J. Ferguson	*Assembly Room* The Square
15 Jul 1771	Philosophical	J. Ferguson	*Assembly Room* The Square
August 1773	Philosophical	J. Arden	*Assembly Room* The Square
29 Mar 1775	The Microcosm	H. Bridges	*Red Lion* Digbeth
Apr 1775	The Microcosm	H. Bridges	*Red Lion* Digbeth
May 1775	The Microcosm	H. Bridges	*Red Lion* Digbeth

Start Date	Description	Lecturer	Venue
4 Mar 1776	Grand Exhibition	G. Katterfelto	*Nag's Head* Snow Hill
29 Oct 1776	Air Lectures	J. Warltire	*Hotel* Temple Street
4 Nov 1776	Air & Chemistry	J. Warltire	*Hotel* Temple Street
4 Aug 1778	Philosophical	Mr. Pitt	*Assbly Room* Cherry Orchard
18 Aug 1778	Philosophical	Mr. Pitt	*Assbly Room* Cherry Orchard
Nov 1778	Philosophical	J. Warltire	*Hotel* Temple Street
8 Dec 1778	Philosophical	J. Warltire	*Hotel* Temple Street
Apr 1779	Philosophical	Mr. Donn	*Hotel* Temple Street
May 1779	Philosophical	Mr. Donn	*Hotel* Temple Street
10 Apr 1780	Chemistry	J. Warltire	*Cooke's* Cherry Street
8 May 1780	Solar Apparatus	J. Warltire	*Cooke's* Cherry Street
13 Nov 1780	Philosophical	J. Warltire	*Cooke's* Cherry Street
Jan 1781	Philosophical	J. Warltire	*Coffee House* Cherry Street
19 Mar 1781	Philosophical	J. Warltire	*Coffee House* Cherry Street
26 Mar 1781	Philosophical	J. Warltire	*Coffee House* Cherry Street
2 Apr 1781	Philosophical	J. Warltire	*Coffee House* Cherry Street
30 Jul 1781	Philosophical	A. Walker	*Assembly Room* The Square
5 Nov 1781	Transparent Orrery	A. Walker	*New Theatre* New Street
24 Dec 1781	Human Propagation	Dr. Graham	*Assembly Room* The Square
12 Jun 1782	Chemistry	J. Warltire	*Coffee House* Cherry Street
3 Jul 1782	Chemistry	J. Warltire	*Coffee House* Cherry Street
21 Oct 1782	Philosophical	Dr. Moyle	*Hotel* Temple Street
19 May 1783	Transparent Orrery	A. Walker	*New Theatre* New Street
2 Jun 1783	Transparent Orrery	A. Walker	*New Theatre* New Street
7 Nov 1785	Philosophical	J. Booth	*Red Lion* Digbeth
19 Dec 1785	Philosophical	J. Booth	*Assembly Room* The Square
21 Feb 1786	Philosophical	J. Booth	*Payton's Auction Room* Upper Priory
8 Nov 1786	Astrotheatron	Mr. Long	*New Theatre* New Street
27 Nov 1786	Astrotheatron	Mr. Long	*New Theatre* New Street
Dec 1788	Philosophical	J. Booth	*Apollo* Deritend
7 Jan 1789	Philosophical	J. Booth	*Mackorkell's* Temple Row
18 Feb 1789	Philosophical	J. Booth	*Mackorkell's* Temple Row
10 Feb 1790	Philosophical	J. Warltire	*Coffee Pot* Cherry Street
6 Mar 1790	Philosophical	J. Warltire	*Coffee Pot* Cherry Street
15 Feb 1791	Philosophical	Mr. Burton	*Shakespeare Tavern* New St.
3 Oct 1791	Philosophical	Mr. Banks	*New Public Rooms* New St.
9 Apr 1792	Lectures/Exhibition	G. Katterfelto	32 New Street
16 Apr 1792	Lectures/Exhibition	G. Katterfelto	32 New Street
23 Apr 1792	Lectures/Exhibition	G. Katterfelto	32 New Street
Jul 1792	Lectures/Exhibition	G. Katterfelto	32 New Street
Aug 1792	Lectures/Exhibition	G. Katterfelto	32 New Street
Sep 1792	Lectures/Exhibition	G. Katterfelto	32 New Street
Oct 1792	Lectures/Exhibition	G. Katterfelto	32 New Street
Nov 1792	Lectures/Exhibition	G. Katterfelto	32 New Street
Dec 1792	Lectures/Exhibition	G. Katterfelto	32 New Street
Jan 1793	Lectures/Exhibition	G. Katterfelto	32 New Street
Feb 1793	Lectures/Exhibition	G. Katterfelto	32 New Street
Mar 1793	Lectures/Exhibition	G. Katterfelto	32 New Street

Source: *Aris' Gazette*

APPENDIX B
Astronomical Items In The MacFarlane Collection Inspected By James Watt In 1756.

There were 14 boxes (marked A – I, K – L) of astronomical equipment shipped to Britain accompanied by a detailed inventory (in French) which included some large items as listed below plus a large number of lenses, eyepieces, various components for mounts, compasses, and books.

Month Clock by Graham 1731.

4 ft Brass 102 inch Arch by Sisson 1730 (this is a half scale copy of the

Greenwich 8 inch iron quadrant by Graham).

12 inch Gregorian telescope by Short 1743.

4 ft Paste Board refractor c1730.

Two Small Refractors with micrometers c1730.

5 ft Transit Instrument c1730.

4 ft combined Transit and equal Altitude instrument c1730.

18 inch Meridian telescope by Graham 1731.

5 ft Portable Zenith Sector c1730.

Horizontal Reflecting Zenith Sector (24 inch Sector) by Martel 1755.

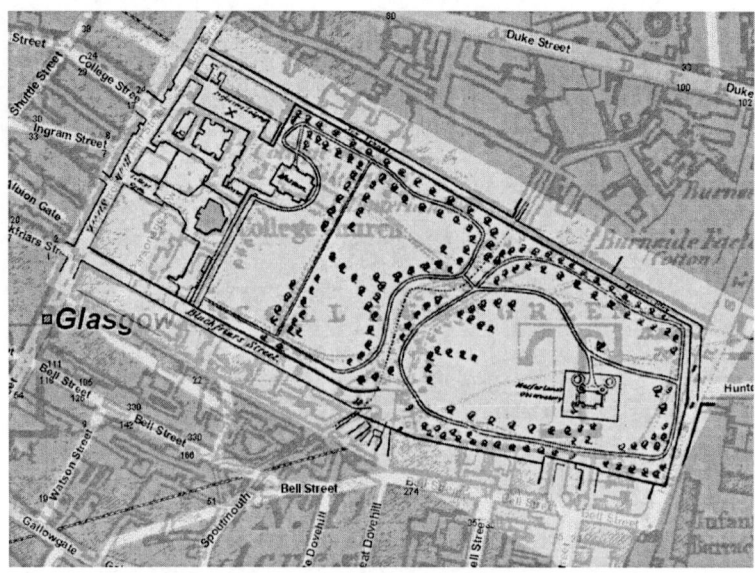

Fig. 52 Position of MacFarlane Observatory, Glasgow
(University of Glasgow)

APPENDIX C
Lunatick Astronomy Dramatic Lectures/Events

Date	Event	Venue
06-Jan-02	Dramatic Lecture	Soho House Museum Birmingham
08-Mar to 01-Sep-02	Exhibition	Soho House Museum Birmingham
02-Nov-02	Dramatic Lecture	SHA/ Soho House Museum Birmingham
01-Jun-03	Dramatic Lecture	Soho House Museum Birmingham
30-Aug-03	Dramatic Lecture	SHA/ Manchester Astronomical Society
15-Mar-04	Dramatic Lecture	Erasmus Darwin House Lichfield
22-Apr-04	Dramatic Lecture	Swansea Astronomical Society
04-May-04	Dramatic Lecture	Soho House Museum Birmingham
11-Jun-04	Dramatic Lecture	Bristol Astronomical Society
15-Sep-04	Dramatic Lecture	Wycombe Astronomical Society
11-Nov-04	Dramatic Lecture	Ancestral Rescue Club Drayton Bassett
07-Jan-05	Dramatic Lecture	West Bromwich Local History Society
01-Feb-05	Dramatic Lecture	Hinckley & District Astronomical Society
16-Jun-05 x2	Dramatic Lecture	Sutton Coldfield U3A Middleton Hall
11-Sep-05	Dramatic Lecture	Soho House Museum Birmingham
21-Oct-05	Dramatic Lecture	Tiverton Astronomy Society
24-May-06	Dramatic Lecture	Halesowen U3A
13-Nov-06	Dramatic Lecture	Staffordshire & Worcester Canal Society
14-Nov-06	Dramatic Lecture	Leicester Astronomical Society
09-Feb-07	Dramatic Lecture	Hall Green U3A, Birmingham
07-Jun-07	Dramatic Lecture	Dorridge U3A
12-Sep-07	Dramatic Lecture	Warwickshire District U3A
02-Oct-07	Dramatic Lecture	Redditch U3A
21-Feb-08	Dramatic Lecture	Henley & Beaudessert Civic Society
11-Mar-08	Dramatic Lecture	Streetly In-Betweens
03-Apr-08	Dramatic Lecture	Bath U3A
07-Apr-08	Dramatic Lecture	Sheffield Astronomical Society
11-Apr-08	Dramatic Lecture	Weston Super Mare U3A
03-Jul-08	Dramatic Lecture	Evesham & District U3A
13-Sep-08	Dramatic Lecture	PyCon 2008 Conference Birmingham
04-Oct-08	Dramatic Lecture	SHA Conference Birmingham & Midland Inst.
27-Oct-08	Dramatic Lecture	Wolverhampton Astronomical Society
12-Jan-09	Dramatic Lecture	Sutton Coldfield U3A
15-Jan-09	Dramatic Lecture	Gorway (Walsall) Probus
26-Jan-09	Dramatic Lecture	Sutton Coldfield U3A
17-Feb-09	Dramatic Lecture	Dudley Canal Trust
25-Mar-09	Dramatic Lecture	Wolverhampton Ladies' Luncheon Group
14-Apr-09	Dramatic Lecture	Birmingham & District Local History Assn.
01-Jun-09	Dramatic Lecture	Bradford Astronomical Society
05-Jun-09	Dramatic Lecture	The Herschel Group, Bath
14-Jun-09	Display	Birmingham History Fair
03-Jul-09	History Paper	University of Birmingham
30-Jul-09	Dramatic Lecture	Balsall Heath Local History Society
23-Sept-09	Dramatic Lecture	Marston GreenAgers, Birmingham
01-Oct-09	Dramatic Lecture	Cardiff Astronomical Society
16-Oct-09	Dramatic Lecture	Cromwell WI, Leicester
26-Oct-09	Dramatic Lecture	Balsall Common U3A
09-Feb-10	Dramatic Lecture	Countesthorpe U3A, Leicester
09-Mar-10	Dramatic Lecture	Evesham Civic Society

Acknowledgements

It is important to acknowledge a number of people regarding this research. Val Loggie the former Curator of *Soho House Museum* has since 1999 been a source of constant encouragement to continue the research and to place it in print, and her advice on Boulton and the 18th century has been invaluable. I am additionally grateful to the members of staff at *Soho House Museum* who have allowed me access to survey the roof and provide facilities to present details of the research to the general public. Shena Mason is another Boulton expert whose book about his daughter Anne is an important work, Shena has given useful advice and checked references. Fiona Tait of *Birmingham City Archives* has gone beyond her duty to cross-check references and documents for me, and volunteered her time to proof read the second edition. The members of staff at *Birmingham City Archives* have been of great assistance (and patience) retrieving countless documents for me over the years. I am grateful to *Birmingham Museum and Art Gallery* for granting me access to the Phillp book and supplying images. The members of staff in the *Science Department of Birmingham Central Library* have assisted in finding copies of science papers from the 18th century. Technical advice on steam engines, instrument making and railway gauges was supplied by Michael Horne who is an Assistant UK Coordinator for *The Planetary Society* and founder member of the *Society for the History of Astronomy*. Dr. Desmond King-Hele's work on Erasmus Darwin has been invaluable and saved me a great deal of time. The members of staff at the *Hunterian Museum* at the *University of Glasgow* have provided invaluable information on James Watt and were so courteous to allow me access to their store of scientific instruments. I am indebted to John Doran the curator at the *Avery Historical Museum* at *Soho Foundry* for allowing me to study the collection of books relating to James Watt held at the museum. The members of staff at *Kidderminster Library* showed great patience in unlocking book cabinets so I could trace details of James Ferguson's Kidderminster lectures. I am grateful for the access given to me by *University College London* to their library. I would like to thank Rory Cook at the *Documentation Centre of the Science Museum* in London, and to the *Museum for the History of Science* at *Oxford University* for supplying details of 18th century telescopes and other instruments. Barbara Fogarty of the *'Lunaticks'* group was kind enough to send me details of documents relating to astronomy she found in the Boulton archive. I am very grateful to Peter Hingley of the *Royal Astronomical Society* for his advice, assistance and allowing me access to the Herschel Collection of documents. Dr. Allan Chapman of *Wadham College Oxford* and Emily Winterburn of the *Royal Observatory Greenwich* have encouraged me to place my research in print and offered useful advice and support. *The Society for the History of Astronomy* has been very supportive in allowing me to present details of my researches. Professor Carl Chinn of the *University of Birmingham* has offered much support and encouragement on this and several other projects which has spurred me to continue. Finally I must acknowledge the late Dr. Mary Bruck (1925-2008) formerly of the *University of Edinburgh* who reviewed my original paper and gave invaluable advice.

LUNATICK ASTRONOMY NOTES AND REFERENCES

The Birmingham City Archives are located in the Central Library Chamberlain Square and those relating to this paper are filed under: Matthew Boulton Papers, James Watt Papers, Matthew Robinson Boulton Papers, and the Boulton & Watt Collection. The Herschel Papers are located at the Royal Astronomical Society Library.

Introduction

1. Schofield, Robert E. *The Lunar Society of Birmingham* Oxford University Press 1963. p17. The term *Lunatick* was apparently coined by Erasmus Darwin which is a typical remark for his sense of humour; the term was also used, according to Mary Ann Galton (Samuel Galton Jr.'s daughter), by the Galton's Butler at their home Barr Hall.
2. Schofield, Robert E. *The Lunar Society of Birmingham* Oxford University Press 1963. p.17.
3. Statement made by James Keir in a memoir of Matthew Boulton written in 1809. The original document is in the Birmingham Assay Office a Photostat is in Birmingham Archives located in a bound volume of *Collection of Letters written by Boulton, Watt and Others.* (MS 3782).
4. Robinson, Eric, *The Lunar Society and the Improvement of Scientific Instruments I* Annals of Science Vol.12 1956 pp.296-304; *The Lunar Society and the Improvement of Scientific Instruments II* Annals of Science Vol.13 pp.1-8 1957.

Chapter 1 A Town of Practical Science

1. William Camden (1551-1623) wrote of his own visit to Birmingham 1584 stating that the town was "swarming with inhabitants and echoing with noise of anvils".
2. For a good detailed history of Birmingham's growth in this period refer to Gill, Conrad *History of Birmingham Volume 1* Oxford University 1952.
3. Aris' Gazette 2nd August 1742.
4. Aris' Gazette 22nd November 1742. The exhibition had been at Worcester and Warwick before arriving in Birmingham.
5. Aris' Gazette 26th March 1744.
6. Aris' Gazette 23rd March, 7th May, and 25th June 1744.
7. Aris' Gazette 20th October, 27th October, 3rd November and 29th December 1746, and 5th January 1747.
8. Aris' Gazette 15th August 1748. The inn may have well been the *Woolpack* an old establishment that had been frequented by Dr. Samuel Johnson and John Baskerville. In the 19th century Sir Rowland Hill the founder of the Penny Post (and son of Thomas Wright Hill) established the *Society for Literary Improvement* at the inn.
9. Aris' Gazette 17th & 24th April 1749. Science problem answers printed 22nd November 1742. Arden may well have been known to John Whitehurst.
10. Aris' Gazette 31st July, & 21st August 1749.
11. Aris' Gazette 4th May 1752.
12. Aris' Gazette 1st July 1754.

13. Aris' Gazette 6th , 21st, & 28th April, 5th May, 2nd & 23rd August, 6th, 13th, 20th & 27th September 1756.
14. Aris' Gazette 6th September 1756.

Chapter 2 Lunar Society's Astronomical Beginnings

1. Boulton's financial situation improved even further in 1764 when Luke Robinson junior died unmarried and thus his inheritance went to Ann and thus to Matthew Boulton, thus he had come into Luke Robinson Senior's entire fortune.
2. Notebooks preserved in the Birmingham Central Library Archive in the Matthew Boulton Papers (MS 3782/12/108). Each book is numbered although the numbers were added by Matthew's son Matthew Robinson Boulton when he began to catalogue his father's papers. Notes in the books are not strictly chronological with notation being made from front to back and back to front.
3. Matthew Boulton's Notebook No. 1, 1751-1759 (MS 3782/12/108) p.7.
4. Matthew Boulton's Notebook No. 1 1751-1759 (MS 3782/12/108) p.25.
5. Aris' Gazette 23rd August 1747 tickets for the Martin Lectures could be purchased through Thomas Aris at his bookshop. Aris' Gazette 23rd April 1750 the lectures ran from 23rd to 28th April.
6. Aris' Gazette 15th January 1755.
7. Hutton, William *The History of the Hutton Family* Mose Haughton 1799 p.176
8. Advertisement in Aris' Birmingham Gazette 24th August 1761. This was for a second series held at the same facility. The Square was renamed 'Old Square' and today the central oval section (enlarged) forms a traffic island at the top of Corporation Street.
9. Hill, Joseph & Dent, Robert K., Memorials of the Old Square Achilles Taylor, Old Square Birmingham 1897 pp.25-33, 84-85.
10. Millburn, John R., *James Ferguson's Lecture Tour 1771* Annals of Science Vol.40 p.404.
11. Advertisements in Aris' Birmingham Gazette 24th August 1761, 7th September 1761 and letter from James Ferguson to Mr. Balfour dated 4th December 1761 where he stated he gave four series of lectures (National Library of Scotland MS 581/569).
12. James Ferguson letter to Mr. Balfour (National Library of Scotland (MS 581/569).
13. Bill of sale for books from Pearson & Aris to Matthew Boulton dated March 1762. The bill covers purchases from March 1759 with Ferguson's book being bought on 23rd September 1761. (MS 3782/6/189/30). This book along with Ferguson's *Tables and Tracts* 1767, and Lectures *on Several Subjects* 1773 were sold at auction by Christie's of London on 12th December 1986 (Auction Catalogue *Books from the Library of Matthew Boulton and his family* p.31 Lot 77). Thomas Pearson and Thomas Aris also published Aris' Gazette.
14. This book was sold at auction by Christie's of London on 12th December 1986 (Auction Catalogue *Books from the Library of Matthew Boulton and his family* p.35 Lot 100).
15. Matthew Boulton's Notebook Number 4 1765 (MS 3782/12/108) p.9.

16. Matthew Boulton's diary 1768 in October he notes: Bought of Dollond 3 wedges of glass to explain refraction £1 11s 6d. (Matthew Boulton Diaries MS 3782/12/107).

17. Schofield, Robert E. *The Lunar Society of Birmingham* Oxford University Press 1963. p.106 and letter from John Michell to Matthew Boulton dated 5th July 1758 (M1 248).

18. Cadbury, Paul *Bicentenary of the Lunar Society of Birmingham* London University Press 1966 p.67.

19. Geike, A. *A Memoir of John Michell* Cambridge 1918 p.165.

20. *The Assay Office at Birmingham Part 1 Its Foundation* Cornish Brothers Birmingham 1936 p.31. The Crown and Anchor has a very special place in the history of silver, originally London had the only Assay Office and used the stamp mark of a Leopard. Birmingham and Sheffield would need different marks and thus they chose the Crown for Sheffield and the Anchor for Birmingham.

21. Letter from Peter Dollond to William Small dated 5th September 1765 (MS 3782/12/23/54).

22. For a very detailed analysis of Watt's early life see Hills, Richard. *James Watt Volume 1 His Time in Scotland 1736-1774* Landmark Publishing 2002.

23. Bryden, D.J. *The Jamaican Observatories of Colin Campbell, FRS and Alexander MacFarlane FRS* Notes and Records of the Royal Society Vol. 24 1970.

24. Letter from James Watt to his father dated 25th October 1756. (4/11.2).

25. A minute from a meeting held at Glasgow University of 2nd December 1756 states: "A precept was signed to pay James Watt Five pounds Sterling for cleaning and refitting the instruments from Jamaica". Referenced in Muirhead, James Patrick *The Origin and Progress of the Mechanical Inventions of James Watt* John Murray 1858 Vol. 1 p.xxxi.

26. James Watt's Scrapbook 1757-1760, note dated 17th November 1759. A copy of this book is in Birmingham City Archives.

27. James Watt's Scrapbook 1757-1760 A copy of this book is in Birmingham City Archives.

28. Letter from James Lind to James Watt dated 9th December 1765 (MS 3219/4/56).

29. Letter from James Lind to James Watt dated 9th December 1765 (MS 3219/4/56).

30. Letter from James Lind to James Watt dated 19th January 1768 (MS 3219/4/56).

31. Letter from James Lind to James Watt dated 29th October 1768 (MS 3219/4/56).

32. Letter from James Watt to Dr. William Small 31st January 1770 (MS 3782/12/76).

33. Letter from William Small to James Watt dated February 1771 (MS 3782/12/76) referenced in Schofield, Robert E. *The Lunar Society of Birmingham* Oxford University Press 1963. p.106.

34. Letter from Matthew Boulton to Duke of Richmond dated 8, April 1772 (Letter book F MS 3782/1/38) this also appears in Taylor, E.G.R. *The Mathematical Practitioners of Hanoverian England* Cambridge University Press for The Institute of Navigation 1966 p.247.

35. Matthew Boulton's Notebook Number 6 1768-1775 (MS 3782/12/108) p.63 and letter from Alexandre Tournant to Matthew Boulton dated 29th May 1774 (MS 3782/12/24).

36. Letter from Matthew Boulton to Duke of Richmond dated 8, April 1772 (letter book F MS 3782/1/38) referenced in Schofield, Robert E. *The Lunar Society of Birmingham* Oxford University Press 1963. p.106 and Smiles, Samuel. *Lives of the Engineers Boulton & Watt* John Murray 1878 p.124.

37. Daumas, Maurice translated by Holbrook, Mary Scientific *Instruments of the 17th & 18th Centuries and their Makers* Portman Books 1989 p.268.

38. Letter from Alexander Tournant to William Small dated 13th April 1775 referenced in Robinson, Eric *The Lunar Society and the Improvement of Scientific Instruments II* Annals of Science Vol.13 March 1957 p.7. I have been unable to locate the original letter.

39. Letter from William Small to James Watt 19th October 1771 (Bound book of letters relating to William Small, Shelf 407).

40. Joseph Priestley states in 'Memoirs ed 1806 p.93' that "Mr. Parker of Fleet Street very generously supplied me with every instrument that I wanted in glass". Such instruments were purchased through Boulton's agent John Wyatt.

41. James Watt's Scrapbook 1757-1760 21st September 1759 sold a 2ft telescope to William Cross.

42. Prices of telescopes as listed by Benjamin Martin in 1762. These and other telescope prices are mentioned in Court, Thomas & Rohr, Moritz von, *A History of the Development of the Telescope from about 1675 to 1830 Based on Documents in the Court Collection.* Transactions of the Optical Society Vol. xxx, No.5, 1928-29.Cambridge University Press. 1929 pp.230-1.

43. Catalogue and price list of Benjamin Martin annotated by James Watt (MS 3219/4/298/5).

44. Letter from Samuel Small to James Watt dated 27th January 1773. (MS 3782/12/76). Schofield, Robert E. *The Lunar Society of Birmingham* Oxford University Press 1963. p.107 and Robinson, Eric, *The Lunar Society and the Improvement of Scientific Instruments II* from Annals of Science March 1957.

45. Letter from James Watt to William Small 13th March 1773 (Bound book of letters relating to William Small, Shelf 407).

46. James Watt's Journal Parts 5-7 1771-1774 bound copy in Birmingham Archives pp.40-41.

47. Letter from Samuel Small to James Watt dated 29th March 1774. Schofield, Robert E. *The Lunar Society of Birmingham* Oxford University Press 1963. p.107 (Bound book of letters relating to William Small, Shelf 407).

48. Letter from William Small to James Watt dated 27th October 1773 (MS 3782/12/76).

49. The prismatic micrometer is described in a letter from James Watt to William Small dated 24th November 1772 and referenced in Smiles, Samuel, *Lives of the Engineers Boulton and Watt* John Murray 1878 p.103.

50. Philosophical Transactions of the Royal Society Volume 67 XXXV pp.789-798. Read June 19th 1777.

51. Philosophical Transactions of the Royal Society Volume 67 XXXVI pp.799-815. Read December 18th 1777.

52. Letter from William Irvine to James Watt dated 2nd July 1778

(MS 3219/4/85).

53. Letter from Patrick Wilson to James Watt dated 2nd July 1778 (MS 3219/4/85).

54. Letter from Gilbert Hamilton to James Watt dated 3rd July 1778 (MS 3219/4/18/7).

55. Letter from William Irvine to James Watt dated 11th March 1775 (MS 3219/4/78/9).

56. Letter from Patrick Wilson to James Watt dated 10th March 1775 (MS 3219/4/78/8).

57. Philosophical Transactions of the Royal Society Volume 69 XXVII pp.414-431. Read March 25th 1779.

58. Muirhead, James Patrick, *The Origin and Progress of the Mechanical Inventions of James Watt* John Murray 1858 Vol 1 pp.cxxxi – cxivi The whole chapter gives an excellent history of the micrometer.

59. Letter from Samuel Small to James Watt dated 29th March 1774. Schofield, Robert E. *The Lunar Society of Birmingham* Oxford University Press 1963. p.107 (Bound book of letters relating to William Small, Shelf 407).

60. Matthew Boulton's Notebook Number 13 c1776-1777 (MS 3782/12/108) p.174.

61. Erasmus Darwin's Commonplace Book pp.80, 84 Darwin Museum, Down House, Kent.

62. Ibid.

63. Keir, James Translator *A Dictionary of Chemistry Containing the Theory and Practice* T. Cadell & P. Elmsly 1771.

64. Schofield, Robert E., *The Lunar Society of Birmingham* Oxford University Press pp.172-4.

65. Ibid.

Chapter 3 The Soho Observatory

1. Letter from John Whitehurst to Matthew Boulton dated 24th May 1771. Whitehurst writes: "I am glad to find you are arrived safe at Soho, and happy in the company of my ingenious friend and philosopher Mr. Ferguson." Whitehurst also writes that he will visit while Ferguson is at Soho. (MS 3782/2/81/13).

2. Tickets for the lectures were available from Aris' Gazette or from Ferguson at Mr. Blockley's Bull Street. Aris' Gazette 27th May 1771. Sketchley's Birmingham Directory 1770 p.38.

3. James Ferguson's Syllabus printed 1771 (Royal Observatory Edinburgh).

4. Advertisement in Aris' Birmingham Gazette 27th May 1771.

5. *Cornish's Stranger's Guide To Birmingham* Cornish Brothers Birmingham 1839 and subsequent editions mention the meridian line. Hutton's *History of Birmingham* and *History of Birmingham* Volume 1 by Conrad Gill do not mention Ferguson and the meridian line. The tower and spire of St. Martin's remain the oldest parts of the existing church most which was rebuilt in 1855.

6. A note from James Ferguson to Matthew Boulton dated 15th July 1771 asks if he could add a link to a chain for him (MS 3782/1/20/13).

7. Letter from James Ferguson to Matthew Boulton dated 10th August 1771 (MS 3782/12/23/213).

8. Hill, Thomas Wright & Hill, Matthew Davenport *Remains of the Late Thomas Wright Hill FRAS* Private printing 1859 p.88.

9. Hill, Thomas Wright & Hill, Matthew Davenport *Remains of the Late Thomas Wright Hill FRAS* Private printing 1859 p.89.

10. Letter from James Ferguson to James Beresford dated 5th November 1771. Ferguson adds that 'Thomas the clockmaker' seems to have forgotten to see Howell. (Am 82-12 held at Princeton University USA).

11. Millburn, John R., *James Ferguson's Lecture Tour of the English Midlands 1771* Annals of Science Vol.42 p.408.

12. Dates calculated from Ferguson's letter dated Saturday 10th August where he states he will be finished on Thursday (15th).

13. Hill, Thomas Wright & Hill, Matthew Davenport *Remains of the Late Thomas Wright Hill FRAS* Private printing 1859 p.116.

14. Millburn, John R., *James Ferguson's Lecture Tour of the English Midlands 1771* Annals of Science Vol.42 p.409.

15. Letter from Josiah Wedgwood to Thomas Bentley dated 13th October 1771 referenced in *Letters of Josiah Wedgwood 1771 -1780* Didsbury, Manchester 1903 p.46.

16. Matthew Boulton's Notebook Number 8 1772 (MS 3782/12/108) p.9.

17. Ibid.

18. Ibid.

19. Ibid.

20. Ibid.

21. Matthew Boulton's Notebook Number 8 1772 (MS 3782/12/108) p.8.

22. Ibid. & Matthew Boulton's Notebook Number 8 1772 (MS 3782/12/108) p.15.

23. Donnelly, Marian Card. *A Short History of Observatories* University of Oregon Press 1973. pp.29 – 51.

24. Matthew Boulton's Notebook Number 8 1772 (MS 3782/12/108) p.25.

25. Letter from Matthew Boulton to Alexander Dalrymple dated 25th July 1772 (MS 3782/12/24).

26. Letter from Alexander Dalrymple to Matthew Boulton (Boulton & Fothergill) dated 21st September 1774 (MS 3782/12/24).

27. Letter from Matthew Boulton to James Watt dated 9th March 1775 (MS 3219/4/66).

28. Letter from J. Ramsden to Matthew Boulton 24, August 1775 (MS 3782/12/24/43). This letter has been incorrectly classified as from 'J. Barnode' although the handwriting is clearly Jesse Ramsden's. This telescope was sold by Sotheby's at auction on 20th March 2003 as lot 73 for £9,600 (Sotheby's Catalogue for *The James Watt Sale Art & Science* London 2003 with printed sales prices).

29. Letter from Matthew Boulton to James Watt dated 26th January 1775 (MS 3219/4/66).

30. Letter from James Watt to Matthew Boulton dated 24th January 1775 (MS 3782/12/76).

31. Letter from Jesse Ramsden to Matthew Boulton 24th August 1775 (MS 3782/12/24/43).

32. Ibid.

33. Letter from Alexander Aubert to Boulton and Fothergill, 25th September 1778 (MS 3782/1/27/20).

34. Album of drawings by John Phillp in the possession of Birmingham Museum & Art Gallery currently undergoing conservation dated 1798; the illustration of the building is signed and dated 1796.
35. Illustrated and described by Thomas Bugge in *Observationes Astronomique* 1784.
36. Donnelly, Marian Card. *A Short History of Observatories* University of Oregon Press 1973. p.51.
37. Thomas Bugge's Log Book of his trip to England in 1777 in possession of Det Kongelige Bibliotek, Copenhagen, Denmark.
38. James Wyatt carried out numerous work for Boulton on Soho House over the years as can be seen from the Matthew Boulton Papers.
39. Letter from Matthew Boulton to James Watt dated 25th June 1778 Boulton comments that he showed Watt's daughter the eclipse.
    (MS 3147/3/459) The partial solar eclipse was visible from Birmingham and easily from Soho House. The eclipse began at 15.53 reaching mid eclipse at 16.37 and ending at 17.41. The Birmingham newspaper 'Aris' Birmingham Gazette' made no mention of the eclipse.
40. Dickinson, H.W. & Jenkins R. *James Watt and the Steam Engine*. Encore Editions 1981. pp.34 – 41.
41. Note by Matthew Boulton to Aubert attached to letter from Alexander Aubert to Boulton and Fothergill, 25th September 1778 (MS 3782/1/27/20).
42. Letter From Alexander Aubert to Matthew Boulton 3rd Oct 1778 (MS 3782/12/24/212).
43. *A Catalogue of the Very Valuable and Extensive Collection of Astronomical and Mathematical Instruments of the Late Alexander Aubert Esq of Highbury House Islington which will be sold at auction by Leigh and S. Sotheby July 21st to 24th July 1806*. p.16 item 254.
44. List of data connecting Matthew Boulton with astronomy forming part of an application for funds for Soho House Museum to purchase a globe of the Moon by John Russell. Soho House Museum.
45. Oxford Dictionary of National Biography H.S. Torrens.
46. Oxford Dictionary of National Biography E. I. Carlyle, rev. Anita McConnell.
47. Aris' Gazette 2nd July 1781.
48. Aris' Gazette 19th May 1783. The New Theatre was built in 1774 and had seating for 2000.. William Hutton described it as one of the best theatres in Britain. It was renamed the *Theatre Royal* in 1807.

Chapter 4 Fireballs and Firebrands

1. Letter from Patrick Wilson to James Watt dated 10th March 1775 (MS 3219/4/78/8).
2. Letter from William Irvine to James Watt dated 11th March 1775 (MS 3219/4/78/9).
3. Letter from Patrick Wilson to James Watt dated 22nd September 1783 (MS 3219/4/92/17).
4. Ibid.
5. Letter from Gilbert Hamilton to James Watt dated 21st August 1783 (MS 3219/4/19/23).

6. Letter from Richard Lovell Edgeworth to Sir Joseph Banks dated 5th September 1783 which appears in Philosophical Transactions of the Royal Society Volume 74 XII p.118.

7. Butler, Marilyn *Maria Edgeworth A Literary Biography* Oxford University Press 1972 pp.192-3, 444.

8. Letter from Nevil Maskelyne to William Herschel dated 24th November 1783. William Herschel Papers W1-13/1.

9. Philosophical Transactions of the Royal Society Volume 74 XVII pp.201-232.

10. Royal Society Archives –Letters and Papers Decade VIII No.102. Referenced in King-Hele, Desmond *Erasmus Darwin A Life of Unequalled Achievement* DLM 1999 p.190.

11. King-Hele, Desmond *Erasmus Darwin A Life of Unequalled Achievement* DLM 1999 p.270.

12. Letter from James Lind to Matthew Boulton dated 28th September 1783 (MS 3147/3/378/13).

13. Ibid.

14. Letter from Miss De Luc to James Watt dated 21st November 1787 (MS 3219/4/99/22).

15. Matthew Boulton's Notebook Number 13 c1778 (MS 3782/12/108) pp.121-122. James Simons worked between 1771-1794 located at Marylebone Street, London.

16. James Simons Instrument maker account rendered to William Withering 9th, October 1782. Withering Papers Royal Society of Medicine, London.

17. Letter from Joseph Priestley to William Eden, 5th December 1771, American Philosophical Society, Philadelphia referenced in Schofield Robert E. editor *A Scientific Autobiography of Joseph Priestley 1733-1804* M.I.T. Press Cambridge Massachusetts p.97.

18. Letter from Joseph Priestley to Sir Joseph Banks 10th December 1771. Weld, C.R. *History of the Royal Society Vol. 2* John W. Parker 1848 pp.56-57 referenced in Schofield Robert E. editor *A Scientific Autobiography of Joseph Priestley 1733-1804* M.I.T. Press Cambridge Massachusetts p.98.

19. Priestley, Joseph *The Memoirs of Dr. Joseph Priestley to the Year 1795* J. Johnson 1806 pp.61-67.

20. Birmingham Assay Office *Matthew Boulton's Otaheite Medal* Cornish Brothers Birmingham 1926 pp.1-9.

21. Letter from Joseph Priestley to William Herschel dated 12th August, 1780. Dreyer, J.L.E. editor *Scientific Papers of Sir William Herschel* Vol. 1 p.xxxi Royal Society & Royal Astronomical Society 1912 referenced in Schofield, Robert E. editor *A Scientific Autobiography of Joseph Priestley 1733-1804* M.I.T. Press Cambridge Massachusetts 1966 p.186.

22. Letter from John Michell to Joseph Priestley dated 21st January 1781 printed in Lubbock, C.A. *The Herschel Chronicle* Cambridge University Press 1933 p.91.

23. Letter from John Michell to William Herschel in 1781 partly printed in Lubbock, C.A. *The Herschel Chronicle* Cambridge University Press 1933. p.91.

24. Herschel, William *Selected Scientific Papers of William Herschel* Royal Society & Royal Astronomical Society 1912 pp.78-81.

25. Spoken in a sermon by Joseph Priestley on November 5th 1785 quoted in Bird, Vivian *The Priestley Riots 1791 and the Lunar Society* BMI 1991 p.22.

26. The dinner was held at Dadley's Hotel in Temple Row, Birmingham on July 14th, 1791. The chairman of the dinner was actually James Keir with 80 other diners.

27. The Old Meeting House had previously been burnt down in the Tory Sachaverell riot of 1715.

28. Hill, Thomas Wright & Hill, Matthew Davenport Remains of Thomas Wright Hill Private printing 1859 pp.117-118.

29. Letter from Joseph Priestley to William Withering dated October 2nd 1792 as referenced in Schofield, Robert E., A Scientific Autobiography of Joseph Priestley 1733-1804 MIT Press 1966 p.264.

30. Bird, Vivian. *The Priestley Riots 1791 and The Lunar Society* Birmingham and Midland Institute 1991 pp.44-53.

31. King George III expressed that he felt Priestley got what he deserved, and the Marquis of Buckingham wrote to Lord Grenville stating that he was not sorry "even for this excess" Bird, Vivian. *The Priestley Riots 1791 and The Lunar Society* Birmingham and Midland Institute 1991 p.54. Given that Birmingham has been by tradition a rebellious town, the riots do seem surprising although there is evidence that some pre-planning had taken place including a flyer campaign. Thomas Wright Hill's wife always felt that High Church officials were involved in organising the riots. The trials of those very few arrested resulted in only 4 guilty verdicts two of which resulted in executions.

Chapter 5 William Herschel's Lunatick Friends

1. Caroline performed as first principle with William as fifth on 15th April 1778. Mrs. John Herschel editor *Memoirs of Caroline Herschel* John Murray 1879 p.42.

2. Letter from William Herschel to Matthew Boulton dated 2nd December 1786 (237/146).

3. Letter from Sir Joseph Banks to Matthew Boulton dated 14th February 1787 (MS 3782/12/56/6).

4. Letter from Matthew Boulton to Matthew Robinson Boulton dated 30th July 1787 (MS 3782/13/36/9).

5. Drawings of proposed modifications to Soho House by John Rawsthorne (MS 1682).

6. William Herschel's Diary pp.4-5 Herschel Papers W7-15/1.

7. William Herschel's Diary p.5 Herschel Papers W7-15/1.

8. William Herschel's Diary pp.8-9 Herschel Papers W7-15/1.

9. William Herschel's Diary pp.9-10 Herschel Papers W7-15/1.

10. William Herschel's Diary p.15 Herschel Papers W7-15/1.

11. William Herschel's Diary p.24 Herschel Papers W7-15/1.

12. William Herschel's Diary p.73 Herschel Papers W7-15/1.

13. William Herschel's Diary p.97 Herschel Papers W7-15/1.

14. Letter from Joseph Black to James Watt dated 7th August 1788 referenced in Robinson, Eric & McKie Douglas *Partners in Science* Harvard University 1970 p.169.

15. William Herschel's Diary p.136 Herschel Papers W7-15/1. A year later Hershel bought the telescope for £30. William Herschel's Diary in the Herschel Papers W0007-0015-0001 p.123.
16. William Herschel's Diary p.162 Herschel Papers W7-15/1.
17. William Herschel's Diary p.170 Herschel Papers W7-15/1.
18. Letter from William Herschel to Matthew Boulton dated November 1792 (MS 3782/12/37/209).
19. The Copying Company was a partnership of Matthew Boulton, James Watt and James Keir.
20. Letter from William Herschel to Matthew Boulton dated July 1793 (MS 3782/12/38/129).
21. Letter from Charles Lincoln to James Watt dated 12th March 1793 (MS 3219/4/111/19) A bill for the telescope (and a couple of other items) is also in the archive; the telescope cost £7 7 shillings. (MS 3219/4/111/20).
22. Scribbled note by Matthew Boulton (MS 3219/4/111/28).
23. Letter from Charles Lincoln to James Watt dated 27th April 1793 (MS 3219/4/111/31).
24. The ships he saw were of the homeward bound West India Fleet. The next day he observed the Russian fleet off Falmouth. Matthew Boulton's Note book Number 24 p.12 (MS 3782/12/108/24).
25. Letter from Charles Lincoln to James Watt dated 12th March 1793 (MS 3219/4/111/19) the pocket telescope is listed on the bill as 'Military' and cost £1 12 shillings (MS 3219/4/111/20). This telescope was sold by Sotheby's at auction on 20th March 2003 as Lot 383 being part of James Watt Junior's items for £216 (Sotheby's Catalogue for *The James Watt Sale Art & Science* London 2003 with printed sales prices).
26. Letter from William Herschel to Matthew Boulton dated 5th March 1794 (MS 3782/12/38).

Chapter 6 Astronomical Witnesses

1. Oxford Dictionary of National Biography H.S. Torrens.
2. Letter from James Lind to James Watt dated 21st June 1793 (MS 3219/4/112/2).
3. William Herschel's Diary pp.12-14, 21-31 Herschel Papers W7-15/1.
4. William Herschel's Diary pp.12, 15-18 Herschel Papers W7-15/1.
5. William Herschel's Diary p.46 Herschel Papers W7-15/1.
6. William Herschel's Diary pp.99-105 Herschel Papers W7-15/1.
7. William Herschel's Diary p.14 Herschel Papers W7-15/1.
8. William Herschel's Diary p.186-194 Herschel Papers W7-15/1.
9. William Herschel's Diary p.14 Herschel Papers W7-15/1.
10. William Herschel's Diary p.186-194 Herschel Papers W7-15/1.
11. William Herschel's Diary p.225 Herschel Papers W7-15/1.
12. William Herschel's Diary p.186-194 Herschel Papers W7-15/1. It was in Redruth that Boulton and Watt feared violent action from Cornish mine owners over the patent indeed on one occasion Boulton hoped that soldiers might be deployed to quell the unrest.

13. Dickinson, H.W. & Jenkins R. *James Watt and the Steam Engine* Encore Editions 1989 p.325.

14. Letter from James Watt Junior to William Herschel dated 26th November 1796. (Letter book 19 item 111).

15. Extract from the shorthand notes by Mr. Gurnsey of the *Minutes of the Proceedings of the trial Boulton and Watt versus Hornblower and Mabberley in the Court of Common Pleas Guild Hall held by a Special Jury on 16th December 1796, Right Honourable Lord Chief Justice Eyre presiding* pp. (MS 3219/4/227).

16. Extract from the shorthand notes by Mr. Gurnsey of the *Minutes of the Proceedings of the trial Boulton and Watt versus Hornblower and Mabberley in the Court of Common Pleas Guild Hall held by a Special Jury on 16th December 1796, Right Honourable Lord Chief Justice Eyre presiding* pp.91-92 (MS 3219/4/227).

17. Extract from the shorthand notes by Mr. Gurnsey of the *Minutes of the Proceedings of the trial Boulton and Watt versus Hornblower and Mabberley in the Court of Common Pleas Guild Hall held by a Special Jury on 16th December 1796, Right Honourable Lord Chief Justice Eyre presiding* pp.98-106 (MS 3219/4/227).

18. Extract from the shorthand notes by Mr. Gurnsey of the *Minutes of the Proceedings of the trial Boulton and Watt versus Hornblower and Mabberley in the Court of Common Pleas Guild Hall held by a Special Jury on 16th December 1796, Right Honourable Lord Chief Justice Eyre presiding* pp.106-7 (MS 3219/4/227).

19. Matthew Boulton's Notebook number 14 1778 p.27 (MS 3782/12/108/14).

Chapter 7 Matthew Boulton and the Instrument Makers

1. Letter from Boulton & Fothergill to Peter & John Dollond dated 17th October 1772 (Letter book F MS 3782/1/38).

2. Letter from Matthew Boulton to Peter Dollond dated 19th September 1776 (Letter book H MS 3782/1/39).

3. Letter from Matthew Boulton to Jesse Ramsden dated 19th September 1776 (Letter book H MS 3782/1/39).

4. Letter from Matthew Boulton to Peter Dollond dated 12th October 1776 & Letter from Matthew Boulton to Jesse Ramsden dated 12th October 1776 (Letter book H MS 3782/1/39).

5. Letter from Matthew Boulton to Jesse Ramsden dated 17th February 1777 & letter from Matthew Boulton to Peter Dollond dated 21st June 1777 (Letter book H MS 3782/1/39).

6. Letter from Matthew Boulton to Peter Dollond dated 3rd July 1778 (Letter book H MS 3782/1/39).

7. Letter from P.P. Burdett to Matthew Boulton dated 15th September 1777 (MS 3782/12/24/122). This letter is catalogued as being from 'S.P. Burdett' and has been correctly identified as being from Peter P. Burdett by Val Loggi who brought to my attention the astronomical and optical content on 16th June 2008.

8. Letter from William Ludlam to Matthew Boulton dated 5th May 1775 (MS 3782/12/24/35) Ludlam explains to Boulton that details of his

clock may be found in the book 'Astronomical Observations at Cambridge' printed for Cadell in the Strand 1769. Boulton obtained a copy which was sold at auction by Christie's of London on 12th December 1986 (Auction Catalogue *Books from the Library of Matthew Boulton and his family* p.44 Lot 135). George Donisthorpe of Birmingham was the brother-in-law of William Ludlow. He supplied a great deal of instruments including barometers, thermometers, clocks etc to Boulton; he also cleaned and repaired Boulton's clocks as can be seen from his bill in the archives (MS 3782/6/190/132).

9. Letter from William Law & Neuman to Boulton & Fothergill dated 27th December 1777 (MS 3782/1/26/24/107).

10. Letter from Joseph Poli to Matthew Boulton dated 4th July 1786 (MS 3147/3/518/1).

11. Letters from Sarah Ramsden to Matthew Boulton dated 16th August 1787 (MS 3147/3/518/6). 10th October 1787 (MS 3147/3/518/7a).

12. Steam engine for the King of Naples was erected at Carditelli near Capua and was used for pumping water for fields near one of the King's Palaces. Invoice of engine supplied to the King of Naples April-May 1788 (MS 3147/3/518/10).

13. Letter from Jesse Ramsden to Matthew Boulton dated 29th December 1781 (MS 3782/12/26/122).

14. British Patent No. 1316 18th May 1782, sealed April 4th 1782.

15. Court, Thomas & Rohr, Moritz von, *A History of the Development of the Telescope from about 1675 to 1830 Based on Documents in the Court Collection.* Transactions of the Optical Society Vol xxx, No.5, 1928-29. Cambridge University Press. 1929 p.242 and King, Henry C., *The History of the Telescope.* Dover Publications 2003 p.161.

16. George Adams of Fleet Street, John Jones of Holburn, and J.H. Tiedemann of Stuttgart as mentioned in Court, Thomas & Rohr, Moritz von, *A History of the Development of the Telescope from about 1675 to 1830 Based on Documents in the Court Collection.* Transactions of the Optical Society Vol xxx, No.5, 1928-29. Cambridge University Press. 1929 pp.242-243.

17. Letter from Jesse Ramsden to Matthew Boulton dated 11th January 1782 (MS 3782/12/27/3).

18. Robinson, Eric. *An Exhibition to Commemorate the Bicentenary of the Lunar Society of Birmingham* Exhibition Booklet Birmingham Museum and Art Gallery 1966. pp.63-64.

19. Letter from John Hodges on behalf of Matthew Boulton to Jesse Ramsden dated 5th October 1782, bill for 18 pieces of plated metal at a cost of £20 13/- plus carriage of 2/2. (Soho Manufactory Letter Book N (MS 3782/2/14).

20. Letter from John Hodges for Matthew Boulton to Jesse Ramsden dated 7th June 1784 (Soho Manufactory Letter Book N MS 3782/2/14).

21. King, Henry C. *The History of the Telescope* Dover Publications 2003. pp.168-169. The Dunsink instrument was ordered in 1785, Ramsden died in 1800 and the work was completed by Matthew Berge and installed in 1808.

22. Letter from Jesse Ramsden to Matthew Boulton dated 6th September 1786 (MS 3782/12/31/108).

23. King, Henry C. *The History of the Telescope* Dover Publications 2003. pp.167-168.
24. Letter from Jesse Ramsden to Matthew Boulton dated 27th January 1777 (MS 3782/12/24/93).
25. Ibid.
26. Letter from Jesse Ramsden to Matthew Boulton dated 30th October 1777 (MS 3782/12/24/130).
27. Letter from Jesse Ramsden to Matthew Boulton dated 27th January 1777 (MS 3782/12/24/93).
28. Letter from Jesse Ramsden to Matthew Boulton dated 30th October 1777 (MS 3782/12/24/130).
29. Probably Thomas Weeks of Titchbourne Street London.
30. Matthew Boulton Diary 1793 (MS 3782/12/107/21).
31. Craven, Maxwell *John Whiteurst of Derby Clockmaker and Scientist 1713-88* Mayfield Books Ashbourne Derbyshire 1996 pp.107-8.
32. Letter from Joseph Wright to Dr. Bates dated 12th September 1772 as referenced in Bemrose,W. *The Life & Works of Joseph Wright* ARA London 1885 p.12.
33. Derby Mercury 13th April 1764.
34. Josiah Wedgwood wrote to Mr. Bentley on 11th September 1767 regarding his enquiry to Whitehurst regarding an astronomical clock but having had no response he was somewhat irked, although he does comment that it is likely because Whitehurst is working on his Geological Thesis. Referenced in Schofield, Robert E. *The Lunar Society of Birmingham* Oxford University Press 1963 p.105.
35. Nathanial Hill of The Sun & Globe Chancellery Lane London. According to various sources he had stopped supplying globes in 1764 with William Palmer taking over the business so it is possible that Ferguson obtained the globe from Palmer which still had the name *N. Hill* printed on it. This might explain the error in size as the globe may have been of a standard size from old stock. Ferguson himself had made globes but he seems to have sold this business to Benjamin Martin which makes it a surprise as to why Ferguson did not go to Martin, although there may have been a falling out as the two were at times rivals on the lecture circuit.
36. Goodison, Nicholas *Ormolu The Work of Matthew Boulton* Phaidon 1974 p.111.
37. Letter from Matthew Boulton to John Whitehurst dated 23rd February 1771 (Letter Book E MS 3782/1/9).
38. Letter from Boulton & Fothergill to John Whitehurst dated 26th August 1771 (Letter Book E MS 3782/1/9). The drawing would have been given to Boulton's workman while Ferguson was staying at Soho House in 1771.
39. Letter from Matthew Boulton to John Whitehurst dated December 1771 (Letter Book E MS 3782/1/9).
40. Matthew Boulton's Notebook Number 8 1772 (MS 3782/12/108) p.11.
41. Letter from John Fothergill to Matthew Boulton 2nd March 1775. Referenced in Goodison, Nicholas *Ormolu The Work of Matthew Boulton* Phaidon 1974. Original letter has not been traced.

42. Letter from William Porter to Boulton & Fothergill dated 29th October 1779. Referenced in Goodison, Nicholas *Ormolu The Work of Matthew Boulton* Phaidon 1974. Original letter has not been traced.

43. Letter from John Fothergill to Matthew Boulton dated 16th December 1779 (MS 3782/12/60/182).

44. Letter from William Porter & Co. to Matthew Boulton dated 2nd November 1787 (MS 3782/12/32/168).

45. Urania watch stand was sold by Christie & Ansells at Auction on 16th May 1778 for £17 Guineas.

Chapter 8 Boulton's Astronomical Twilight

1. Soho Foundry was bought by W & T Avery Limited in 1895 where weighing scales, bridges etc were produced. The site is now owned by Avery Weigh-Tronix Limited.

2. The Soho House roof was extensively surveyed during restoration. The survey was carried out by the City of Hereford Archaeology Unit in 1992. An inspection of the roof was made by Andrew Lound and Michael Horne in 1999 to examine if the telescope platform was practical.

3. Shaw, Stebbing *The History and Antiquities of Staffordshire* J. Nichols and Son London 1798-1801.p.121.

4. Matthew Boulton's Diary 1800 July entry (MS 3782/12/107/28).

5. Letter from Matthew Boulton to O. Warburg dated 5th February, 1801 (MS 3782/12/46/58). In 1804 Warburg negotiated with Boulton on behalf of the Danish Mint regarding coinage.

6. Letter from Anne Boulton to Matthew Boulton dated 15th January 1801 (MS 3782/13/38/10).

7. Matthew Boulton's Diary 1801. This might be John Webb of 327 Oxford Street, London (MS 3782/12/107/29).

8. William Herschel's Diary pp.253, 284 Herschel Papers W7-15/1

9. Letter from James Watt Jr. to James Watt dated 4th September 1806. Watt Jr. writes: "I hear that Dr. Herschel has been here..." Watt Jr. had been away and arrived back at Soho Foundry on 3rd September. (MS 3219/4/32/25).

10. William Herschel's Diary p.284 Herschel Papers W7-15/1 Letter from John Herschel to Matthew Robinson Boulton dated April 2nd 1829 attests to the visit. (MS 3782/13/22/112).

11. Philosophical Transactions of the Royal Society Volume 74, pp.35-57 1784. Read November 27th 1783.

12. King-Hele, Desmond. *Erasmus Darwin A Life of Unequalled Achievement* DLM 1999.' p.263.

13. Erasmus Darwin's Commonplace Book pp.80, 84. Darwin Museum, Down House, Kent.

14. William Herschel's Diary pp.296 Herschel Papers W7-15/1.

15. Letter from Professor Mylne to James Watt dated 22nd July 1813 (MS 3219/4/290/4).

16. Lubbock, C.A. *The Herschel Chronicle* Cambridge University Press 1933. p.355.

17. Letter from John Herschel to Matthew Robinson Boulton dated April 2nd 1829 attests to the visit. (MS 3782/13/22/112).

Bibliography

Barker, Richard (editor) *Wilkinson Studies* Vols.1 & 2 Merton Priory Press 1991-2.

Bemrose,W. *The Life & Works of Joseph Wright* ARA London 1885.

Bird, Vivian. *The Priestley Riots 1791 and The Lunar Society* Birmingham and Midland Institute 1991.

Birmingham Assay Office *Matthew Boulton's Otaheite Medal* Cornish Brothers Birmingham 1926.

Burton, Anthony *Josiah Wedgwood A Biography* André Deutsch 1976.

Butler, Marilyn *Maria Edgeworth* Oxford University 1972.

Clarke, Desmond *The Ingenious Mr. Edgeworth* Oldbourne 1966.

Clifton, Gloria, *Directory of British Scientific Instrument Makers 1550-1851* Zwemmer/National Maritime Museum 1995.

Court, Thomas & Rohr, Moritz von, *A History of the Development of the Telescope from about 1675 to 1830 Based on Documents in the Court Collection* Transactions of the Optical Society Vol xxx, No.5, 1928-29.Cambridge University 1929.

Craven, Maxwell *John Whitehurst of Derby Clockmaker and Scientist 1713-88* Mayfield Books Ashbourne Derbyshire 1996.

Daumas, Maurice, *Scientific Instruments of the 17th & 18th Centuries and their Makers* Portman Books 1989.

Dick, Malcolm *Joseph Priestley and Birmingham* Brewin Books Studley 2005.

Dickinson, H.W. *Matthew Boulton* Cambridge University Press 1937.

Dickinson, H.W. & Jenkins R. *James Watt and the Steam Engine* Encore 1989.

Donnelly, Marian C. *A Short History of Observatories* University of Oregon 1973

Geike, A. *A Memoir of John Michell* Cambridge Univerity Press 1918.

Goodison, Nicholas *Ormolu The Work of Matthew Boulton* Phaidon 1974.

Griffiths, John *The Third Man (William Murdoch)* Andre Deutsch 1992.

Henderson, E. *Life of James Ferguson F.R.S.* A. Fullerton & Co. 1867.

Herschel, Mrs. John *Memoir and Correspondence of Caroline Herschel* John Murray 1879.

Herschel, William *Selected Scientific Papers of William Herschel* Royal Society & Royal Astronomical Society 1912.

Hill, Joseph & Dent, Robert K. *Memorials of the Old Square* Achilles Taylor Old Square Birmingham 1897.

Hills, Dr. Richard L. *James Watt Volume 1: His Time in Scotland 1736-1774* Landmark Publishing, Ashbourne Derbyshire 2002.

Keir, James Translator *A Dictionary of Chemistry Containing the Theory and Practice* T. Cadell & P. Elmsly 1771.

King, Henry C. *The History of the Telescope.* Dover Publications 2003.

King-Hele, Desmond *Erasmus Darwin A Life of Unequalled Achievement* DLM 1999.

Langford, John A. *A Century of Life in Birmingham 1741-1841* Simpkin & Marshall 1868

Lubbock, C.A. *The Herschel Chronicle.* Cambridge University 1933

Mason, Shena *The Hardwareman's Daughter* Phillimore & Company Chichester 2005.

Millburn, John R. *The London Evening Courses of Benjamin Martin & James Ferguson 18th Century Lecturers on Experimental Philosophy* Annals of Science

Vol.40 1983.

Millburn, John R. *James Ferguson's Lecture Tour of the English Midlands 1771*
Annals of Science Vol.42 1985

Millburn, John R. *Wheelwright of the Heavens The Life & Work of James*
Ferguson Vade-Mecum Press 1988.

Muirhead, James Patrick *The Origin and Progress of the Mechanical Inventions*
of James Watt John Murray 1858.

Peck, T. Whitmore & Wilkinson K. Douglas *William Withering of Birmingham*
John Wright & Sons Bristol 1950.

Phillips, Patricia *The Scientific Lady* Weidenfield & Nicholson 1990.

Priestley, Joseph *The History and Present State of Discoveries Relating to Vision,*
Light, and Colours J. Johnson 1772

Priestley, Joseph *The Memoirs of Dr. Joseph Priestley to the Year 1795* J. Johnson
1806.

Robinson, Eric *The Lunar Society and the Improvement of Scientific Instruments I*
Annals of Science Vol.12 1956

Robinson, Eric *The Lunar Society and the Improvement of Scientific Instruments II*
Annals of Science Vol.13 1957.

Robinson, Eric *An Exhibition to Commemorate the Bicentenary of the Lunar*
Society of Birmingham Birmingham Museum & Art Gallery 1966.

Robinson, Eric & McKie, Douglas (editors) *Partners In Science James Watt &*
Joseph Black Harvard University USA 1970

Schofield, Robert E. *The Lunar Society of Birmingham* Oxford University 1963.

Schofield, Robert E. editor *A Scientific Autobiography of Joseph Priestley 1733-1804*
M.I.T. Press Cambridge Massachusetts 1966.

Shaw, Stebbing *The History and Antiquities of Staffordshire* J. Nichols and Son
London 1798-1801

Sheldon, Peter *The Life & Times of William Withering His Work & Legacy*
Brewin Books Studley 2004.

Taylor, E.G.R., *The Mathematical Practitioners of Hanovarian England*
1714-1840 Cambridge University 1966.

Smiles, Samuel, *Lives of the Engineers Boulton and Watt* John Murray 1878.

Uglow, Jenny *The Lunar Men The Friends Who Made The Future* Faber & Faber
2002.

Weld, C.R. *History of the Royal Society Vol. 2* John W. Parker 1848.

Wolf, A., *A History of Science Technology, and Philosophy in the Eighteenth*
Century George Allen & Unwin 1952.

Additional Sources

Aris' Gazette 1741-1819 held on microfilm at Birmingham City Archives.
A Catalogue of the Very Valuable and Extensive Collection of Astronomical and
Mathematical Instruments of the Late Alexander Aubert Esq of Highbury House
Islington which will be sold at auction by Leigh and S. Sotheby July 21st to 24th July
1806.
Books From the Library of Matthew Boulton & His Family Sales Auction at Christie's
Great Rooms on Friday 12th December 1986.
The James Watt Sale Art & Science London 20th March 2003 Sotheby's Auction.
Oxford Dictionary of National Biography Online
Philosophical Transactions of the Royal Society

For Further Information

Odyssey Class Dramatic Lectures
An Odyssey Class Dramatic Lecture ^{TM} entitled 'Lunatick Astronomy' is available for public or private events. The presentation covers in detail the activities of the Lunar Society's astronomical interests complimented with images, music, sound effects, a wealth of props presented in 18th century costume. Details of this and other Dramatic Lectures may be found at www.odyssey.dial.pipex.com

Soho House Museum
The home of Matthew Boulton and a favourite meeting place of the Lunar Society is open to the general public from Easter to October Tuesday – Sunday Soho House Museum, Soho Avenue off Soho Road, Handsworth, Birmingham B18 5LB   Tel: 0121 554 9122   www.bmag.org.uk

Society for the History of Astronomy
An organization formed to further the work of professional and amateur historians in the field of astronomy. Publishes 'The Antiquarian Astronomer' which contains articles and papers by members on all aspects of astronomy history. For details visit www.shastro.org.uk   .

The Planetary Society
The world's largest space interest group supports space exploration through funding of scientific research and public awareness projects. Publishes a bi-monthly magazine 'The Planetary Report' and through a large network of volunteers stages public events worldwide. For details visit planetary.org .